P9-AGU-989

Advances in Mobile Radio Access Networks

For a listing of recent titles in the *Artech House Mobile Communications Series*, turn to the back of this book.

Advances in Mobile Radio Access Networks

Y. Jay Guo

Artech House
Boston • London
www.artechhouse.com

Library of Congress Cataloging-in-Publication Data
A catalog record for this book is available from the U.S. Library of Congress.

British Library Cataloguing in Publication Data
Guo, Y. Jay
Advances in mobile radio access networks—(Artech House mobile communications library)
1. Mobile communications systems
I. Title
621.3'845

ISBN 1-58053-727-8

Cover design by Yekaterina Ratner

685 Canton Street
Norwood, MA 02062

International Standard Book Number: 1-58053-727-8

10 9 8 7 6 5 4 3 2 1

Contents

Acknowledgments

I would like to thank Mobisphere Ltd. for giving me the privilege to work in the forefront of mobile communications technology, together with 3G industry leaders Siemens and NEC. I would like to express my gratitude to the following R&D leaders and experts for the valuable discussions I had with them on many of the topics presented in the book: Dr. H. Dressler, Dr. N. Endo, Mr. L. Travaglini, Dr. J. Sokat, Dr. M. Schwab, Dr. W. Mohr, Dr. J. Schindler, Dr. M. Kottkamp, Dr. A. Seeger, Dr. M. Breitbach, Mr. M. Wiesen, Dr. H. Kroener, Dr. G. Schnabl, Dr. A. Splett, Dr. J. Mayer, Mr. T. Shimizu, Mr. T. Sato, Mr. K. Tsuji, Mr. K. Tanoue, Dr. Ng Cheng Hock, Dr. G. Hertel, and Mr. H. Singh. It should be pointed out, however, that the material presented in the book represents only my view, not that of any of the three companies.

I am also grateful to some of my former colleagues with whom I worked at Fujitsu on advanced 3G base stations: Mr. S. Vadgama, Mr. Fukuda, Mr. M. Shearme, Dr. M. Davies, Mr. M. Zarri, and Mr. Y. Tanaka. Further, I am indebted to some leading academics whose insight I have benefited from: Professor S. K. Barton, University of Manchester, England; Professor J. Gardiner, University of Bradford, England; and Professor L. Hanzo, University of Southampton, England. I would like to thank Professor F. C. Zheng, Victoria University of Technology, Australia, for his contribution to Chapter 5. Moreover, on behalf of Professor Zheng, I would like to thank Mr. J. C. Campbell of Telstra Research Laboratories, Melbourne, Australia, for his helpful comments on some of the material presented in Chapter 5. My special thanks go to the anonymous reviewers and Artech House for their constructive comments and valuable suggestions that I received in the process of preparing the manuscript.

Last, but not least, I would like to express my gratitude to Clare, Stella, and Charl Guo for their love, inspiration, understanding, and kind support.

Chapter 1

Introduction

This book gives a comprehensive overview of the technologies for the advances of mobile radio access networks. The topics covered include linear transmitters, superconducting filters and cryogenic radio frequency (RF) front head, radio over fiber, software radio base stations, mobile terminal positioning, high speed downlink packet access (HSDPA), multiple antenna systems such as smart antennas and multiple input and multiple output (MIMO) systems, orthogonal frequency division multiplexing (OFDM) systems, IP-based radio access networks (RAN), autonomic networks, and ubiquitous networks. These technologies are aimed at achieving higher data rates, greater coverage and capacity, lower infrastructure cost, ease of operation and maintenance, higher quality of services, and richer user experience. Some of them, such as radio over fiber, HSDPA, and transmit diversity, will become a reality in the near future. Other technologies, such as software radio base stations, smart antennas, multiple input and multiple output (MIMO) systems, IP-based RAN, and autonomic networks, are still regarded by many as research topics for the fourth generation mobile communications networks (4G). It should be noted, however, that these promising technologies are not of pure academic interest. Owing to their compelling advantages, they are being studied by leading mobile network infrastructure vendors and being employed in various field trials. Therefore, they will play major roles in future mobile communications networks. In fact, a few of them have even been adopted already by some local network operators around the globe.

The book is written from the viewpoint of system engineering and is focused mainly on high-level architectural issues. While highlighting the advantages of the advanced technologies, major theoretical and practical problems facing system designers are also discussed. This book aims to serve mobile communications system engineers, researchers, research and development (R&D) managers, and telecom analysts. I strive to strike a balance between theory and implementation, and between technology advance and economics.

1.1 FUTURE EVOLUTION OF MOBILE RADIO ACCESS NETWORKS

With the accelerating deployment of the third generation (3G) mobile communications networks and the debut of various multimode and multimedia mobile terminals, data-centric, high-speed and feature-rich mobile communications services are becoming a reality. Up until now, operators of the mobile communications networks have taken a pragmatic approach to the deployment of the third generation mobile communications networks. The 3G radio access networks deployed in the initial phase are mainly aimed at providing coverage in order to meet the regulatory requirements and the likely demand of early 3G service adopters. As such, the networks have been designed to minimize costs, while providing necessary features to meet the expectation of 3G subscribers. In the meantime, the hardware has been prepared in such a way that future upgrades can be carried out with ease and minimum cost.

It should be pointed out that the technology transition from the second generation (2G) to 3G is fundamentally different from all the earlier transitions. Although 3G does offer much wider bandwidth and therefore higher data rates than 2G, a more profound long-term effect is that it makes it possible to offer new and exciting services and enables subscribers to do things which they have never done before. With these new services, 3G operators can increase their revenues from existing subscribers. By contrast, the benefit of the early transitions was mainly the expansion of the subscriber base. Therefore, the evolution of future mobile radio access networks will probably be driven by services.

As the demand and the variety of 3G services increase, it is expected that the next few years will see the enhancement of the mobile radio communications network infrastructure with more advanced technologies, in both software and hardware. The first example is location-based services. When new applications based on accurate mobile terminal positioning become available, both 2G and 3G networks will start supporting new positioning technologies. The second example is remote radio head and radio over fiber technology. Currently, with the limited penetration of macro cells, it is difficult to offer guaranteed services within large buildings and in underground tunnels, which are normally called blind spots. The remote radio head and radio over fiber technology provides a neat solution to this problem. Its concept is to detach the radio frequency (RF) part, which is referred to as radio head, from the baseband processing part of the base station equipment, which is referred to as base station server. The base station server can be placed at a convenient location and the downlink and uplink signals are sent over optical fibers to and from different radio heads located at the cell sites, typically next to the antennas. In effect, the distributed antennas and radio heads also provide flexible coverage over sectors that may be geographically distant from each other or far away from the base station.

The 3G networks have been facing competition from other hot-spot technologies such as wireless local area networks (WLANs) from the very beginning of the service launch. Undoubtedly, both the UMTS radio access networks (UTRAN) and CDMA2000 1x networks that are two core 3G systems do have the advantages of greater mobility, greater coverage, and high data rate. However their peak data rate may not be as high as that of, say, WLANs. To maintain the competitive advantage of 3G networks, it is expected that the high speed downlink packet access (HSDPA) technology will be introduced to UTRAN in the near future to increase the data rate in the downlink by an order of magnitude. Being often dubbed 3.5G technology, HSDPA will offer a data rate as high as 14 Mbps and greater system capacity, thus enriching user experiences and reducing the cost per packet. Networks that have been up and running can be upgraded by replacing some cards and modules in base stations, which are referred to as node Bs in UTRAN, and in the radio access controllers (RNC). This helps maximize the return on the investment by operators. In the meantime, a new generation of node Bs and RNCs will also be available for network expansion and for late UTRAN adopters. In a similar fashion, the CDMA2000 1xEV-DO (evolution and data optimized) technology is being introduced to American and Asian markets for high speed data services.

Naturally, the time will come when the 3G networks start experiencing capacity problems in urban areas. Then, network operators will require capacity enhancement technologies. One such technology is multiple antennas that include transmit diversity antennas, smart antennas, and multiple input and multiple output (MIMO) systems. By introducing independent radio signal paths and space and time coding, transmit diversity antennas lead to diversity gains at the mobile terminal without much increasing the complexity of the latter. In fact, a two-antenna transmit diversity scheme has been included in the UTRAN standard as an optional feature. Smart antennas technology is aimed at increasing the system capacity by virtue of the antenna array gain and interference reduction. Smart antennas for cellular networks have been around for several years but no mass market take-up has been materialized yet. There are a number of reasons behind the unwillingness of operators to deploy smart antennas technology: the increased hardware cost especially associated with power amplifiers, the difficulty of installing new antennas and coaxial cables on existing sites, and higher maintenance cost. To overcome these difficulties, it is expected that the new generation of smart antennas for future radio access networks will be developed based on the remote radio head, radio over fiber, and linearized RF transmitters. In contrast to transmit diversity and smart antennas, MIMO requires multiple antennas not only at the base station site but also at the mobile terminals. By transmitting a multitude of data streams in parallel, MIMO has the potential of increasing the data rate of a mobile communication system by an order of magnitude. When used for HSDPA, for instance, the peak data rate of UTRAN can be potentially increased to more than 100 Mbps.

Another technology to enhance the data rate in mobile communications networks is the orthogonal frequency division multiplexing (OFDM). The achievable data rate in a wireless system depends strongly on the radio environment, especially the delay spread of the channel that is caused by multiple reflections from surrounding buildings and terrains. OFDM systems offer inherent resilience against the multipath phenomenon. In an OFDM system, the data stream is divided into M parallel substreams and each of them is transmitted over a different carrier. Also, the system is designed so that signals over different carriers are orthogonal; therefore, they do not interfere with each other. As a result, the symbol period is effectively extended by M, thus reducing the relative delay spread of the radio channel with respect to the bit period and allowing the transmission of much higher data rates. Currently, OFDM is being used in high data rate wireless local systems such as wireless local area networks. To increase the data rate of future cellular networks, various OFDM schemes are being considered as the air interface for the fourth generation mobile communications systems.

In a mobile communications network, the radio access nodes (base stations) are managed by radio network controllers (RNC). The current implementation of both UTRAN networks and CDMA2000 1x networks results in highly centralized RNCs. This architecture is prone to catastrophic system failures caused by faults in RNCs and may also lead to unnecessary traffic loads over the expensive transport network. Therefore, the current trend in the mobile communications industry is the development of more distributed architecture. To this end, the next generation RNC will be in the form of user-plane and control-plane servers and the intelligence of base stations will be increased. Also, to take advantage of the ubiquitous Internet protocol (IP) technology and to realize the economy of scale, it is expected that all the servers and radio access nodes will be connected by a common IP network. This new architecture is referred to as IP-based radio access networks (IP-based RAN). IP-based RANs will facilitate the integration and the flexible deployment and radio resource management of the heterogeneous networks. Such integration will also increase the mobility and capacity of the whole mobile communications networks [1].

With the increasing complexity and scale of future mobile communications networks, technologies for network management will become critical for network operators to control the quality of services and operational expenditures. In fact, the management and maintenance of such heterogeneous networks will be a very challenging task. A promising solution is to substantially increase the intelligence level of the network and its elements to enable self-configuration, self-optimization, and self-healing. We call these highly automated networks *autonomic networks* in this book. An autonomic network should be aware of itself, be capable of running itself in an optimal manner, and be self-healing. It should adjust itself to varying circumstances and manage its resources to handle the traffic loads most efficiently. It should be equipped with redundancy in the configurable hardware and with downloadable firmware. When faults happen in the network or when the network is attacked, it should repair the malfunctioning parts and protect itself with minimal or zero human intervention. In an autonomic network, the maximum amount of traffic will be handled with satisfactory quality of services, and minimum human effort and interference will be needed.

When the concept of 3G was introduced in the 1990s, the aim was to build an integrated ubiquitous network so users can access the telecommunications network anywhere and at anytime without awareness of the technology. Unfortunately, this did not happen. The current reality is that UTRAN is being widely deployed in Europe and Asia, and CDMA2000 1x is being deployed in the Americas and Asia. In parallel, a number of IP-based IEEE 802.x family systems are being standardized as wireless extensions to the global Internet. In particular, IEEE 802.11 wireless LANs (WLANs), usually dubbed WiFi, have been widely deployed across the globe for hot-spot services. On one hand, now we do have the technology for body networks, personal networks, vehicle networks, local area networks, and wide area networks. In principle, these networks can be deployed to provide the infrastructure of ubiquitous networks. On the other hand, these networks are based on different access technologies and they do not work together properly. Therefore, they are actually causing confusion and segmentation to the market. From an economic point of view, the interworking issue must be resolved first before introducing any new air interface to future cellular systems. Fortunately, the mobile communications industry has recognized the problem and is starting to work on the interworking of these different systems. This is demonstrated by the effort of 3GPP and 3GPP2 on 3G and wireless LAN interworking, and by the establishment of the new IEEE 802.21 working group for the integration of the IEEE 802.x family systems. It is expected that such an endeavor will help realize the dream of ubiquitous networks.

1.2 OUTLINE OF THE BOOK

A typical mobile radio access network consists of radio access nodes, radio network controllers, and operation and maintenance nodes. The radio access nodes are responsible for connecting the mobile terminals to the radio access network via the air interface and they are normally referred to as base stations in the cellular networks. In the UMTS terrestrial radio access networks (UTRAN), the base stations are called node Bs. The radio network controllers (RNCs) are responsible for the control of radio resources. Their functionalities include radio resource allocation, radio link setup and mobility management, and interfacing with the core network (CN). In the GSM and CDMA2000 1x networks, the radio access controller is termed the base station controller (BSC). As an illustration, Figure 1.1 shows the architecture of UTRAN [2]. The operation and maintenance of the base stations and RNCs in a radio access network are managed by the operation and maintenance nodes. Figure 1.2 shows the architecture of network management in UTRAN. It is seen that the node Bs and RNCs are managed by element managers and these element managers preside on a common management platform, which is normally called the operation and maintenance center (OMC) in UTRAN. It can be seen that the operation and maintenance (O&M) traffic between the node B and the OMC can be sent to each other directly (shown as a dashed line) or routed via the RNC (shown as dotted lines). Finally, the OMC interfaces with the network manager responsible for the whole mobile network. This book provides an overview on the technology advances for mobile radio access networks in all the above areas.

Chapter 2 deals with four base station radio technologies that are independent of any specific air interface, which include the following:

- Linearized transmitters;
- Superconducting filters and cryogenic RF front end;
- Remote radio head and radio over fiber;
- Software radio base stations.

It is well known that the power amplifier (PA) is one of the most important devices in a base station due to its high manufacturing cost and power consumption. There are two major criteria in the design of power amplifiers: linearity and efficiency. With conventional techniques, it is normally difficult to achieve high performance in one aspect without sacrificing the other. A promising solution is to apply the digital adaptive predistortion technique to power efficient nonlinear amplifiers, thus resulting in the linearized transmitter. The superconducting filter and the cryogenic front end play important roles in the uplink. By improving the filtering characteristics and reducing the thermal noise level, they can improve the cell coverage, increase system capacity, and reduce the transmit power of the mobile terminals. Regarding the remote radio head and radio over fiber technology, two types of applications have been envisaged. The first is the reduction of power consumption and PA cost, and the second is the flexible coverage of micro and pico cells as well as base station hoteling. Software radio refers to radio transceivers whose functionalities are largely defined and implemented by software, and therefore they can be reprogrammed to accommodate various physical layer formats and protocols without replacing the hardware. A software radio base station is one implemented using the technology of software radio. The advantages of applying software radio technologies to base stations are the following. First, the future-proof feature and the economy of scale of software radio base stations can reduce the long-term infrastructure cost. Second, software radio base stations make it possible to use compatible infrastructure across different air interface standards, which simplifies the network planning, management, and maintenance, thereby paving the path to autonomic networks. Third, the upgradability of the software radio base stations gives great flexibility to operators in offering new and creative applications and services. In Chapter 2, an architectural level discussion on the above technologies and related technical and economic issues is presented.

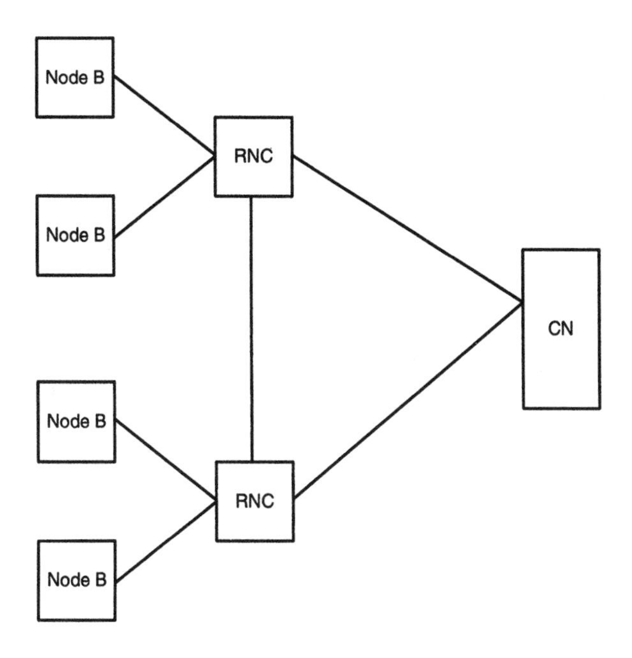

Figure 1.1 An illustration of UTRAN architecture.

Chapter 3 is focused on technologies for mobile terminal positioning. Location-based services are regarded as one of the most important future applications by operators of mobile communications networks. Up until now, four types of positioning techniques have been selected by the 3G Partnership Project (3GPP), which is the standardization body for UTRAN. These include cell ID, assisted global positioning system (A-GPS), observed time difference of arrival (OTDOA), and uplink time difference of arrival (UTDOA). Cell ID is based on the cell coverage area in which the mobile terminal is located. It is the easiest but the least accurate. Assisted GPS is based on the satellite navigation system (GPS) developed by the U.S. Department of Defense. In its operation, each mobile is equipped with a GPS receiver and the positioning is done jointly by both the terminal and the network. Assisted GPS techniques are the most accurate but may not be suited for in-building services. Both OTDOA and UTDOA use time difference measurement and triangulation to locate the mobile terminals in question. The difference between them is that OTDOA is based on the downlink signal and the UTDOA operates on the uplink signal. They offer a good balance between accuracy and robustness. In this chapter, the theory, implementation issues, advantages and disadvantages, and solutions to potential problems of the four positioning techniques are addressed.

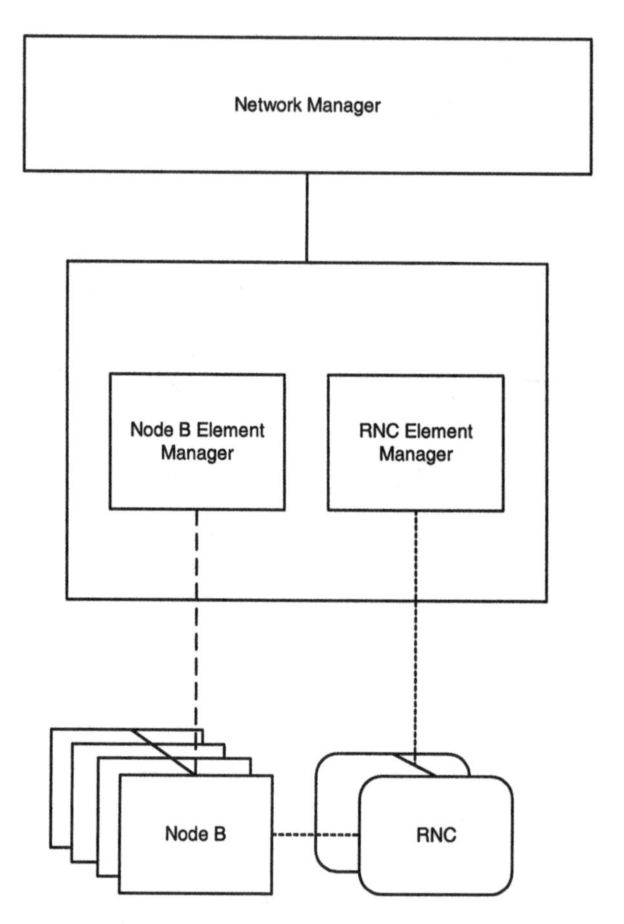

Figure 1.2 The network management architecture for UTRAN.

Chapter 4 offers an introduction to the high-speed downlink packet access (HSDPA) technology. It is expected that, once 3G services are widely adopted, there will be strong demand on both the system capacity and the data rate. This is what is driving the current development of HSDPA in UTRAN and CDMD2000 1xEV-DO. The essence of the HSDPA is the employment of the high-speed downlink shared channel (HS-DSCH), the adaptive modulation and coding scheme, and hybrid automatic retransmission request (ARQ). In Chapter 4, the technical details of HSDPA are given and some practical issues such as network upgrading are discussed.

Chapter 5 is on multiple antenna technology, which is aimed at increasing the system capacity by employing multiple antennas. It is focused on three types of techniques for base stations, including:

- Smart antennas;
- Transmit diversity antennas;
- Multiple input and multiple output (MIMO) antennas.

In Chapter 5, the concepts behind those techniques are explained and the implementation issues are discussed. In particular, the pros and cons of each technique are compared and guidelines for the practical deployment of multiple antenna systems are provided.

Chapter 6 is focused on the orthogonal frequency division multiplexing (OFDM) systems. In this chapter, the operation principle of the OFDM transceivers is introduced first. The practical engineering problems and solutions of the OFDM transmission systems are discussed. Then the theory of an advanced version of OFDM, OFDM/IOTA, is presented. Finally, several OFDM-based proposals for mobile communications systems beyond 3G are described [3, 4].

Chapter 7 is dedicated to the architecture evolution of radio access networks. It is a common view that the future mobile communications network will be a combination of different systems, being integrated by Internet protocol (IP). To make efficient use of network resources, such integration necessarily requires a distributed architecture for handling user-plane traffic, signaling, and radio resource management. In this chapter, some major technologies for IP-based RAN are presented. These include IP transport, mobility management, which covers mobile IP (MIP) and hierarchical mobile IP (HMIP), and distributed network control nodes suited for handling IP traffic. Furthermore, promising architectures that are being considered by the telecom industry are elaborated.

Chapter 8 introduces the concept of autonomic networks. The term autonomic is derived from the body's autonomic nerve system, which controls key functions without conscious awareness or involvement, and the concept of autonomic network is partly borrowed from the world of computing. An autonomic network has three fundamental features: self-awareness, self-optimization, and self-healing. In this chapter, some promising means of realizing autonomic networks, such as artificial intelligence (AI) techniques and the simple network management protocol (SNMP), are discussed. It is shown how the classical AI and distributed AI can be employed to fulfill some network operation and maintenance (O&M) tasks, including performance management and fault management.

Chapter 9 presents a perspective of future mobile communications networks: ubiquitous networks. From the viewpoint of end users, the future mobile communications systems should be ubiquitous and pervasive. This will enable them to access the information network, communicate with each other, and perform various computing tasks anywhere and at any time. The ubiquitous network will connect not only people but also objects. In this chapter, the roles of mobile radio access technologies in ubiquitous networks are discussed. In particular, as candidate components of the future ubiquitous networks, the IEEE 802 family of systems is presented. These include wireless local area networks (WLAN), wireless personal area networks (WPAN), and wireless metropolitan area networks (WMAN). The vision of ubiquitous networks is elaborated by examining the two network convergence paths: the 3G path and the IEEE 802 path. The 3G path is aimed at using the 3G cellular networks, UTRAN and CDMA2000, for wide area coverage and to employ WLANs and WPANs as their extensions for hot-spots, offices, and homes [4, 5]. Using IEEE 802.21 as a unifier, the IEEE 802 path is aimed at building wireless extensions to the Internet by integrating WLANs, WPANs, WMANs and the emerging mobile broadband wireless access (MBWA) network with the Ethernet [6]. The selected examples in this chapter show a glimpse of the future.

References

[1] W. Mohr and W. Konhaiiser, "Access Network Evolution Beyond Third Generation Mobile Communications," *IEEE Communications Magazine*, December 2000, pp.122-133.

[2] H. Holma and A. Toskala, *WCDMA for UMTS*, New York: John Wiley & Sons, 2000.

[3] T. Ohseki, et al., "Proposal of OFDM/MC-CDMA Based Broadband Mobile Communication System," *IEICE Convention*, 2003.

[4] http://www.3gpp.org.

[5] http://www.3gpp2.org.

[6] http://www.ieee802.org.

Chapter 2

Emerging Radio Technologies

A major functionality of the radio access nodes is to connect mobile terminals with the mobile communications network at the physical layer. The format of the radio signal is defined by the specific standard in question. For instance, UTRAN employs the wideband code division multiple access (WCDMA) as the air interface and most of the IEEE 802.11 family systems use the orthogonal frequency division multiplexing (OFDM) scheme (see Chapters 6 and 9). Although a major task in the design of a radio access node or base station is to accommodate the chosen air interface, there are some radio technologies that can be applied to different air interfaces. For example, high performance radio frequency (RF) filters and efficient and linear power amplifiers (PA) can be used in all radio access systems, although the actual implementation must take the specific frequency band into account. In this chapter, four emerging radio technologies that are aimed at solving general problems in base station transceivers are presented, and their applications and potential challenges are discussed. The first technology is called linearized transmitters, whose objective is to increase the efficiency of power amplifiers while maintaining linearity. The second is superconducting filters and cryogenic receiver front end. It applies the high-temperature superconductor and cryogenic technologies in the RF front end and has the potential of reducing mobile terminal transmit power and increasing coverage, system capacity, and spectrum efficiency. The third technology is remote radio heads and radio over fiber, which is to detach the RF part (called the radio head) of the base station transceiver from the baseband part. By connecting the remote radio heads with the baseband part via optical fibers, the required output power of the power amplifiers is reduced, cells at different locations and with different sizes can be flexibly covered, and the base station can be conveniently located. The fourth technology is software radio base stations, which is aimed at realizing most of the base station functionalities in software so that the base stations become upgradable with little or without hardware change. A vision to integrate all these technologies in the future is presented to complete the chapter.

2.1 LINEARIZED TRANSMITTERS

Owing to its high manufacturing cost and power consumption, the power amplifier is one of the most critical devices in a base station or radio access node. There are two major problems in the design of power amplifiers: nonlinearity and low efficiency. Linearity is the ability of an amplifier to deliver output power without distorting the input signal, and the efficiency of an amplifier is defined as the ratio of the output RF power and the total power consumed. Unfortunately, conventional power amplifiers tend to offer either high linearity or high efficiency with the sacrifice of the other. As the importance of the spectral efficiency in mobile communications systems increases, nonconstant envelope digital modulation schemes are being employed in various new air interfaces for different mobile radio access systems [1]. Consequently, the linearity of the radio frequency power amplifiers has become a critical design issue. This issue is particularly important in WCDMA and CDMA2000 1x base stations, where the peak-to-average ratio of the modulated RF signals can vary over a range of, say, 3 to 12 dB. The concern for linearity is primarily due to the stringent restrictions on intermodulation products and out-of-band power emission requirements [2]. Additionally, the amplification of multicarrier signals that is common in the third generation mobile communications systems requires an adequate amplifier linearity in order to avoid significant cross modulation. Furthermore, for spectrum-efficient modulations, amplifier nonlinearity can produce substantial signal distortion and hence increase bit error rates.

Linearity can be achieved, in part, through the use of more linear amplifiers such as class A amplifiers, and by operating the amplifier backed off from the saturation point [2]. In this case, the signal level is confined to the linear region of the amplifier characteristics. However, this approach results in low dc-to-RF conversion efficiency and thus low PA efficiency, which is particularly costly in base station applications, as this leads to huge total power consumption across the network. Furthermore, low dc-to-RF conversion efficiency necessitates high current operating points, which leads to undesired thermal effects.

A viable alternative to using low efficiency linear amplifiers is the application of linearization techniques to more efficient power amplifiers such as class C amplifiers. Among various linearization techniques, adaptive digital predistortion appears to be the most attractive; this has the capability of coping with signals with wide bandwidth and large peak-to-average ratio that is inherent in the third generation and future mobile communications systems. Since the predistortion is implemented digitally, a greater degree of precision can be achieved when computing the predistortion coefficients. Also, unlike analog systems, the imperfection of various components can be tolerated. Furthermore, with the availability of high-speed digital signal processors, adequate million instructions per second (MIP) levels are available to deal with the wideband signals. Last but not least, thanks to the decreasing cost of digital signal processors (DSP), a fully adaptive digital predistortion system is becoming economical.

A fully adaptive digital predistortion system requires a predistortion circuit consisting of a digital predistorter and a look-up table (LUT) to the transmission path and in the feedback path. The predistorter is a nonlinear circuit whose gain response is the inverse of the gain compression of the power amplifier and whose phase response is the negative of that of the power amplifier. For a practical power amplifier, however, the relationship can only be achieved up to the saturation point of the amplifier characteristics (see Figure 2.1). Therefore, the peak-to-average power ratio of the input signal will determine how close to saturation the power amplifier can operate and still behave linearly once the pre-distortion coefficients are applied. For a digital predistortion system, the compensation for the PA nonlinearity can be provided at either the baseband or the IF band. To streamline the system design and to maintain low cost, it is preferable to do it in the baseband. This means that a direct upconversion and down conversion circuit can be integrated into the whole transmitter, which results in the so-called linearized transmitter [3].

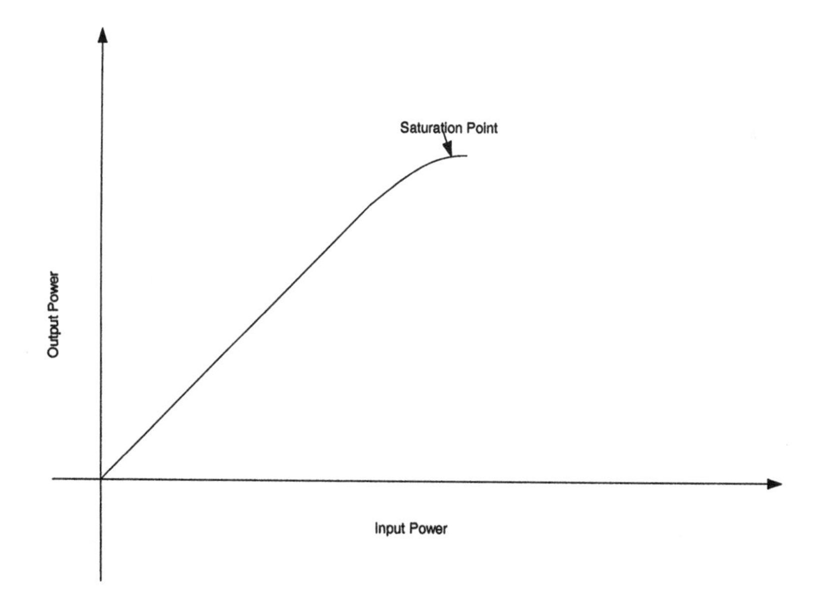

Figure 2.1 Characteristics of a typical power amplifier.

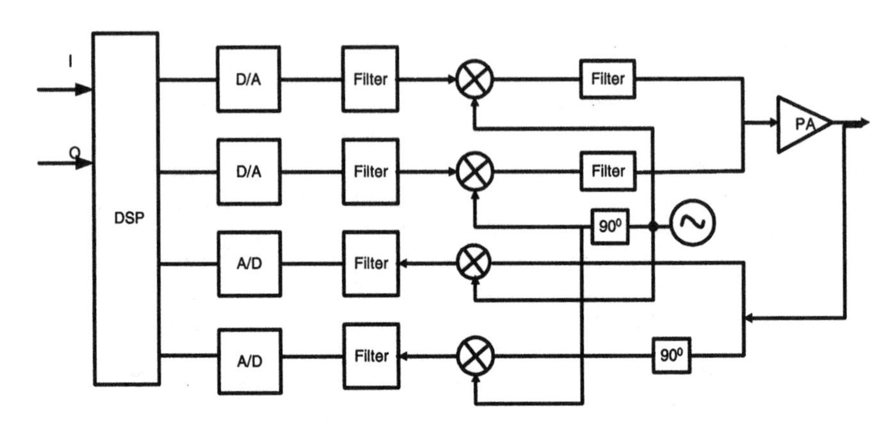

Figure 2.2 The schematic of an adaptive predistortion circuit.

The schematic of a linearized transmitter subsystem is shown in Figure 2.2. The main signal transmission path at the top is comprised of a digital signal processor, a pair of digital to analog converters (D/A), an upconverter, a power amplifier, and some filters. The feedback path at the bottom consists of a down-converter, a pair of analog to digital converters (A/D), and some filters. The digital signal processor (DSP) deals with both the estimation of distortion in the feedback path and the predistortion in the transmission path. It should be noted that the A/Ds, D/As, and the digital signal processors do consume a significant amount of power. To ensure high efficiency of such a system, therefore, care must be taken to ensure that the power consumption in the linearization devices does not offset the efficiency gain achieved by predistortion.

2.2 SUPERCONDUCTING FILTERS AND CRYOGENIC RECEIVER FRONT END

Superconductors are materials that have the ability to conduct electrical current with absolutely no loss of energy. Unfortunately, superconductivity only happens at extremely low temperatures. The lowest temperature in nature is -459°F or -273°C. This temperature is known as "absolute zero" on the Kelvin temperature scale (0K). The science of ultra-low temperatures (<-150°C) is known as "cryogenics" and these temperatures are often called *cryogenic temperatures* [4]. Cryogenic temperatures can be reached using a specially-designed refrigerator, commonly referred to as the *cryocooler*, or by submersing the device to be cooled in a fluid that boils at a low temperature. Liquids that are commonly used to achieve cryogenic temperature are nitrogen, which boils at 77K, and helium, which boils at 4K. Cryocoolers achieve their cooling capability by controlled evaporation of volatile liquids, by controlled expansion of gases confined initially at high pressure, or by alternatively expanding a gas near the area to be cooled and then compressing it at another location in a closed cycle [4].

Until 1986, the only way to make a material superconducting was to reduce its temperature almost all the way to absolute zero. Reaching and sustaining such temperatures are very expensive and difficult. In 1987, a new class of so-called High Temperature Superconductors (HTS) was discovered. They have the same properties as the old superconductors but with one crucial exception: They become superconducting at much higher temperatures. The HTS material was discovered in late 1986 when Müller and Bednorz of IBM's Zurich Lab announced a superconducting oxide at 30K. In 1987, Chu of the University of Houston announced the discovery of a compound, Yttrium Barium Copper Oxide (YBCO), that becomes superconducting at 90K. The next months saw a race for even higher temperatures that produced bismuth compounds (BSCCO) that are superconductive at temperatures up to 110K and thallium compounds (TBCCO) that are superconductive up to 127K [4, 5].

HTS materials above 77K allowed the use of a readily available, low-cost, easy-to-use coolant: liquid nitrogen. Before that, superconductivity required the use of liquid helium, which is both more expensive and more difficult to handle. In addition, these higher superconducting temperatures above 77K made superconductivity possible for the first time using compact, light, and closed-cycle refrigerators. The power consumption of these refrigerators is just a few watts, whereas that for superconductors at 4K is on the order of kilowatts. All these paved the way for the commercial exploitation of the superconducting and cryogenic technologies in radio access networks.

2.2.1 Superconducting Filters

The unique ability of superconductors to conduct electrical current with little or no resistance can be exploited to produce electronic components and systems with greatly increased efficiency and greatly reduced levels of electronic interference and noise. Wireless communications systems utilize filters to prevent unwanted signals from interfering with the wanted ones. A critical filter in a radio access node or base station is the preselect filter, which allows only desired signals in the chosen band to reach the radio receiver. The design of such filters is very challenging. With conventional material, one must balance between the significant insertion loss in the passband and the slope of roll-off at the band boundaries. A filter can be made more selective by adding more filter stages (or poles), but this increases the amount of desired signal that is lost due to energy dissipation in the filter, which is known as insertion loss. Because superconductors are nearly lossless at microwave frequencies, they can be used to make filters with a large number of poles without incurring substantial insertion losses. For a filter designer, superconductors make it possible to provide the closest approximation to a perfect filter, namely, one that allows 100% of the desired signals to pass through the given band and rejects 100% of the unwanted signals in the adjacent bands. In other words, the characteristics of superconducting materials provide a unique ability to produce microwave filters that can simultaneously deliver superior performance with respect to both adjacent band rejection and inband insertion loss. The ability to reject adjacent band interference is important in the base station because such signals can otherwise saturate the amplifier or introduce other nonlinearity in the receiver front end. Further, adjacent band interference introduces distortion into communications channels and reduces system capacity.

In the current UTRAN specification, appropriate guard bands are used to protect against interferences caused by adjacent band signals, but guard bands represent a waste of spectrum resource. Superconducting filters can allow future radio access systems to become much more spectrum efficient. By using extremely sharp adjacent-band rejection filters, mobile communications network operators can partition their available spectrum without introducing significant guard bands or blocked spectrum at the band boundaries, thus maintaining a greater number of channels in the band. Further, high selectivity against interference reduces the effective noise floor, thereby restoring lost capacity and reduced coverage as well as permitting lower mobile terminal transmit power. Given this ability to provide outstanding performance with respect to all the important system parameters, superconducting filters are ideal for use as preselect filters in future base stations. Up until now, the most popular material used for making superconducting filters is YBCO [5, 6], although the use of other material such as TBCCO has also been reported [4].

Most HTS filters employ the conventional metallic planar structures. To achieve the desired superior performance, however, those factors that are normally neglected in the traditional RF circuit design, such as coupling losses and radiation effect, must be considered. Examples of resonating elements of HTS filters include open-ended circular loops, open-ended meander line loops, $\lambda/2$ lines, and π-shaped and symmetrical spiral-shaped lines [6 – 8].

It should be pointed out that it is a challenging task to employ superconducting filters in commercial base stations. One reason is that the connections between the superconducting filter and the normal RF components must be good electrical conductors and poor thermal conductors, but the two properties do not normally go together. Therefore, care must be taken to ensure low losses at the junctions between copper and superconductors. A possible solution is to employ electromagnetic coupling instead of direct physical contact and the other is to place the whole RF front end in the cooler and connect it with the conventional circuit via less temperature-sensitive components. Another challenge facing the design of superconducting filters is that an efficient, low-volume, and low-cost cooling system is needed to keep the system cooled at a certain temperature for years at a time, as these filters would be used on remote radio towers. It is unacceptable for operators to bear the burden of frequent maintenance and for mobile network subscribers to suffer from frequent service interruptions.

2.2.2 Cryogenic Receiver Front End

To further improve the performance of the base station receiver, the cryogenic technology used in superconducting filters can be applied to the entire receiver front end in order to reduce the system noise figure (*NF*). In terms of the signal-to-noise ratio at the input SNR_i and that at the output SNR_o, the noise figure (*NF*) of the receiver front end, or indeed that of any two-port network, is defined as

$$NF = SNR_i - SNR_o \tag{2.1}$$

It can be observed from (2.1) that the noise figure represents the degradation of the signal-to-noise ratio due to the receiver internal noise. The more noise added by the receiver, the lower the signal-to-noise ratio at the input of the demodulator that follows the receiver front end. Cryogenic cooling can significantly decrease RF losses in electronic circuits, thereby reducing the thermal noise, also known as Johnson noise, and the noise figure. In fact, the noise mechanisms intrinsic to a variety of semiconductor transistor designs, such as those used in low-noise amplifiers (LNA), are also temperature-dependent. For example, the noise figure of PHEMT GaAs low-noise amplifiers is known to substantially decrease when operated at cryogenic temperature. When designed properly, the insertion loss of the filters should be no more than 0.2 dB, and the noise figure of the LNAs should be no more than 0.4 dB, so the noise figure of a cryogenic receiver front end could be less than 0.6 dB. This compares favorably with noise figures in the range of 3 to 6 dB in conventional base stations. As will be discussed in Section 2.4, the cryogenic receiver front end can enhance the performance of the analog-to-digital converters (ADC) and increase the receiver selectivity.

2.2.3 Application in CDMA Systems

CDMA systems are particularly vulnerable to interference because users in a cell and its adjacent cells share the same bandwidth simultaneously and they serve as interferers to each other. The power control mechanism causes all mobiles to increase power when interference signals are present in the operation band. These interferers can be inband signals from other users in the cell, or they can be intermodulation products caused by adjacent band signals, typically from either collocated or adjacent cells. The "out-of-band" interference can dramatically reduce coverage and capacity.

These effects have been studied in field trials of superconductor systems by some pioneering HTS filter vendors. Conductus performed one such trial with a suburban CDMA (IS-95) B-band operator that was experiencing interference problems from a competing A-band operator [5]. The goal of the trial was to quantify the capacity improvement that the HTS system could provide by reducing the effects of the interference. Calibrated interference signals were applied at levels comparable to the measured interference at the site. These signals were observed to essentially collapse the coverage area of the site, using only the existing front end. With the superconductor system in place, however, the full coverage and capacity of the cell site were reported to be restored [5].

There are three U.S. companies who have been among the most active in pushing the superconducting filter technology into the commercial cellular market, which include Conductus Inc., ISCO International Inc. [9], and Superconductor Technologies Inc. Their partners on the operator side include Verizon Wireless, Dobson Communications Corp., AllTel Corp., and some Japanese operators [10].

At the moment, there are a number of criteria that superconductor systems must meet in order to gain general industrial acceptance. From a performance perspective, the products must provide a combination of unparalleled high selectivity as well as low noise; these characteristics are the reason for using the technology in the first place. Beyond the electrical performance, however, the superconducting systems should be highly reliable, compact, and field proven, as they are replacing the well-established existing technology. Operators must become comfortable with issues such as cooler reliability, the time required to achieve operational temperature, and the system response to various failure modes or degradation of performance. Last but not least, the price of cryogenic systems must come down dramatically. According to a recent press release [5], a single cryogenic front end could cost $110,000, which is much more than the cost of a whole conventional base station.

As mobile radio access networks evolve from the second generation to the third generation and beyond systems, more capacity will be needed for data-centric applications and more sites will be required. On the other hand, owing to health and environmental concerns, it is becoming increasingly difficult to obtain more sites and there is also strong demand to reduce the transmit power of the mobile terminals. Further, compared with other technologies such as smart antennas, the impact of high temperature superconducting filters and cryogenic receivers on the whole system is relatively small. Therefore, it is foreseeable that there could be a widespread deployment of cryogenic systems in the future once such products become affordable.

2.3 REMOTE RADIO HEAD AND RADIO OVER FIBER

In the conventional base station architecture, the RF parts, including power amplifiers, low noise amplifiers, and up- and downconverters as well as filters and switches, are located in the base station equipment [11]. The radio signals are carried over some RF cables between the base station and the antennas that are normally mounted on the top of the antenna masts. For the downlink, which is from the base station to the mobile antenna, the employment of this configuration means that the power amplifier must provide an output power an extra few decibels higher than what is needed at the antenna in order to compensate for the RF cable losses. Consequently, this results in higher power amplifier cost and higher power consumption. Also, the RF cables carrying high power are costly to manufacture and install and prone to mechanical failures. For the uplink, which is from the antenna to the base station, this configuration leads to unnecessary degradation of the signal-to-noise ratio, thus reducing the coverage area or increasing the terminal power consumption. A solution to these problems is the remote radio head and radio over fiber, which is to place the radio heads away from the rest of the base station, called the base station server, and next to the antenna (see Figure 2.3). The radio resources of a central base station can be organized into several groups either on a fixed basis or dynamically. For the downlink, the intermediate frequency (IF) or baseband signals in each group are electro-optically converted and sent over an optical fiber network. At the other end of the fiber network, the optical signals are converted back to an IF or baseband signal and transmitted via the radio head [12, 13]. For the uplink, it works in a similar way. Typically, the radio head consists of a power amplifier, a low noise amplifier (LNA), a duplexer, an antenna, and some filters. It can also include up- and downconverters and digital-to-analog and analog-to-digital converters.

Compared with the conventional coaxial cable, the optical fiber provides much less attenuation. The attenuation of a typical single mode fiber is about 0.7 dB/km, whereas that for a coaxial cable is about 0.4 dB/m. This means that, using radio over fiber technology, the signals can be carried over much longer distances without employing repeaters. Another benefit of using radio over fiber technology is the large bandwidth. The bandwidth of an optical fiber is thousands of times greater than that of a coaxial cable. Therefore, a single fiber can support not only current mobile communications systems but also systems in the future. A further advantage of the radio over fiber technology is associated with the light weight of optical fibers. This is particularly useful for macro cells employing high towers. If the heavy cables can be replaced with much lighter optical fibers, it would become much easier to introduce such technology as smart antennas. As will be discussed in Chapter 5, one major concern regarding the employment of smart antennas is the implication of introducing a large number of RF cables to feed the antennas, as the RF cables will reduce the reliability and increase the cost and maintenance complexity of the radio access nodes. Another major concern in this respect is the cost of employing multiple power amplifiers. With reduced power consumption and cost of power amplifiers, the total cost of the smart antenna system can be reduced significantly.

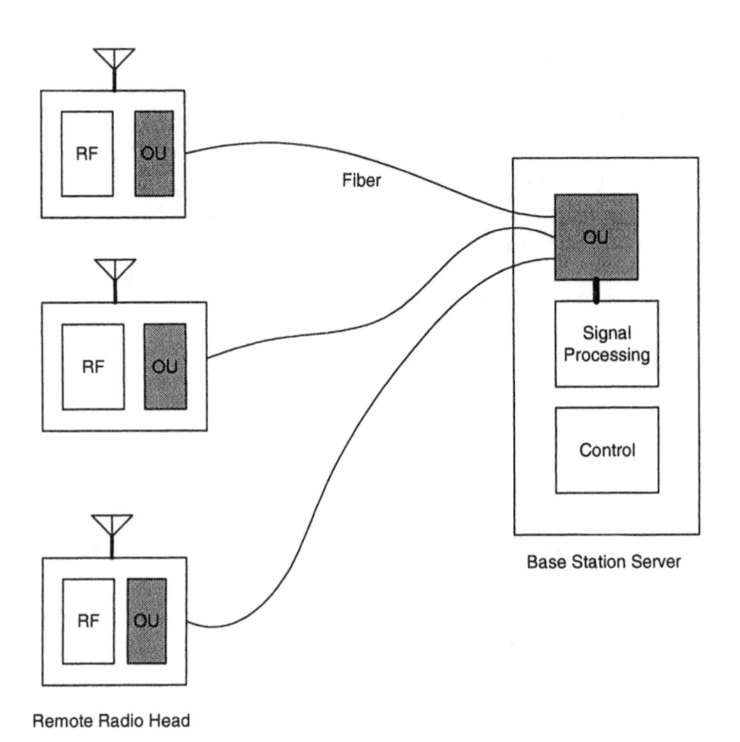

Figure 2.3 Illustration of connecting remote radio heads and base station server with optical fiber. OU refers to optical unit.

Another advantage of the remote RF head technology is its economical coverage for blind spots and large buildings. Conventional cellular technology divides the geographical areas to be covered into cells and the communications between the users in a cell and the network are dealt with by a base station located at the cell site. In some circumstances, however, such an arrangement may not be economical. In a large multistoried building, for instance, the penetration loss of the radio wave through floors and walls makes it difficult to provide coverage from an outdoor cell site. It is much more economical and easier to further divide the building into smaller areas called pico cells (a large room or a few adjacent small rooms) and serve them with slim radio heads distributed within the building, and then the baseband processing is done centrally in a base station server. Basically, the remote radio head technology offers an alternative to pico or micro base stations. This naturally leads to another application of the remote RF head: base station farming or hoteling.

The concept of base station farming is to employ distributed remote radio heads connected to a central processing server, usually located at the cheapest or most convenient place for the network operator. This configuration reduces the footprint of the equipment that must be deployed at each antenna, since a typical remote antenna and radio head are much more compact than even a compact pico base station. Furthermore, operation and maintenance of a single server farm in the processing hotel are likely to be much cheaper for the mobile network operator than the conventional case where a base station is deployed at each antenna site. A further advantage of a hotel configuration is the ability to dynamically balance user load across the server cluster, thus taking advantage of the statistical multiplexing gain to reduce the overall server capacity. For example, consider a system with some antennas in a downtown financial area and others in a suburban residential area. During working hours, the controller assigns a greater number of server nodes to the downtown radio heads. Once workers begin to go home at the end of the day, the system can dynamically migrate capacity over to the suburban radio heads. As a result, the total capital expenditure for operators is reduced.

To facilitate the deployment of remote radio heads, a new industrial specification, common public radio interface (CPRI), has been published [14]. CPRI was created by a consortium of leading mobile infrastructure vendors including Siemens and NEC. In CPRI, a base station is divided into two parts: the radio equipment (RE) and the radio equipment controller (REC) (see Figure 2.4). The radio equipment part includes analog and radio frequency functions such as filtering, amplification, carrier modulation, frequency conversion, and digital-to-analog and analog-to-digital conversion. The REC part is responsible for baseband signal processing, interfacing with the radio network controller, and management and control of the base station. CPRI defines the serial digital interface between the RE part and REC part, in which the user traffic data, the control and management messaging, and the synchronization signal are time-multiplexed. Also, the physical link between the RE and the REC parts via both electrical cables and optical fibers is specified to accommodate the radio over fiber technology.

In order to provide the flexibility to both base station vendors and operators, CPRI defines three reference configurations. The first one is point-to-point link between a single RE and single REC. The second is a multiple of point-to-point links between a single RE and a single REC, and the third is a multiple of point-to-point links between a single REC and multiple REs. The first and the second configurations offer the possibility of locating the RF part of the conventional base station away from the rest of the equipment. The third configuration provides the possibility for several REs to share the baseband processing, control, and management functionalities and the interface with the radio network controller.

2.4 SOFTWARE RADIO BASE STATIONS

The term "software radio" was coined by Joseph Mitola III, a pioneer of software radio, in the early 1990s [15]. At that time, a typical radio consisted of about 80% hardware and 20% software. Mitola suggested that a software radio should have the opposite hardware and software composition, with more software content than hardware content. Generally speaking, software radio refers to radio transceivers whose functionalities are largely defined and implemented by software, and therefore they can be reprogrammed to accommodate various physical layer formats and protocols without replacing the hardware. An ideal software radio should have two distinct features: software upgradability and hardware reconfigurability. The former refers to the fact that, owing to the wide bandwidth offered by the hardware platform and the powerful software, the radio can communicate with not only different systems of existing standards but also future systems of new standards by virtue of software upgrading. The latter refers to the fact that the hardware can be reconfigured on the fly to deal with the chosen air interface more efficiently. Obviously, a software radio base station is one implemented using the technology of software radio. Compared with a peer-to-peer software radio and software radio terminals, a software radio base station does not need to change its functionalities as frequently. On the other hand, it should have the capability of adapting to different network interfaces.

For mobile communications, the driver on software radio is more on the terminal side than on the infrastructure side. This is due to the fact that it is the terminal that needs to roam between different mobile communications networks of different air interfaces, to adapt to the networks available at the specific geographical location, and to choose the one that meets its need best. However, there are some significant advantages in applying software radio technologies to base stations as given in the following.

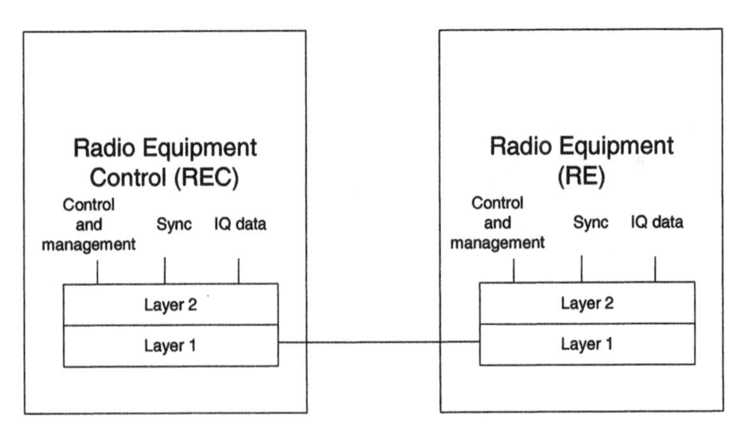

Figure 2.4 The CPRI architecture.

First, the future-proof feature and the economy of scale of software radio base stations can reduce the infrastructure cost of mobile communications network operators. The development cost of the software radio platform can be shared among different markets and projects. This reduces the nonrecurring engineering cost associated with hardware development of digital transceivers to a single development project for multiple market segments. Also, the adoption of a common digital transceiver architecture allows software components supporting one market segment to be reused in another. This reduces vendors' overall cost of software development. Given the fact that older generation systems will coexist with new ones over a long period of time, the economical benefit is potentially great. For instance, a software radio base station for 3G systems can support future air interfaces by upgrading the software and replacing some digital signal processors (DSPs).

Second, software radio base stations make it possible to use compatible infrastructure across different air interface standards, which simplifies the network planning, management and maintenance, thereby paving the path to autonomic networks (see Chapter 8). This is of great importance to operators. In fact, software radio will lead to reduced amount of equipment in the field due to multi-radio support. Since a common set of inventory for a single platform can be utilized for multiple markets, the installation and support cost can be significantly reduced. The base station can be easily scaled up or down from pico to macro nodes according to its needs. It is foreseeable that, equipped with software radio base stations and with a common radio resource manager, an operator can optimize the setting of every base station to minimize interference, ease congestion, and increase data rates, thus achieving higher customer satisfaction and higher profit.

Third, the upgradability of the software radio base stations gives great flexibility to operators in offering new and creative applications and services. Another advantage is that the time to market for the future air interface supported by the platform will be reduced. This is because the dependency of the software development on the hardware development is removed and software reuse enables fast turnaround of applications.

2.4.1 Hardware Architecture

A software radio consists of four essential elements: (1) analog/RF interface and down/upconversion to/from some intermediate frequency (IF), (2) analog-to-digital and digital-to-analog conversion, (3) digital signal processing, and (4) baseband network interface. In an ideal software radio, the functionality of all four elements should be reconfigurable and controllable through software.

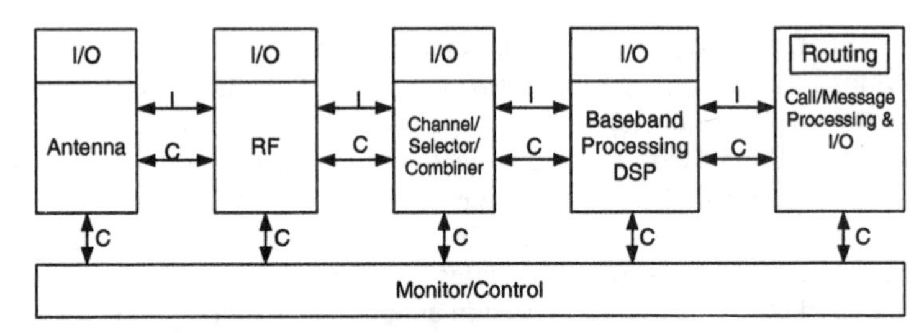

Figure 2.5 The hardware reference model of a software radio base station.

Figure 2.5 shows the base station reference model defined by the software defined radio (SDF) forum [16]. Admittedly, it is fundamentally similar to the current base station architecture used by most infrastructure vendors. A major difference, however, lies in the fact that wideband antenna and RF front end are employed to bring in and send out signals of broad bandwidth, possibly consisting of a sum of narrower band signals of different modulation schemes defined by different standards. The channel selection and combination block is responsible for multiplexing and demultiplexing signals for different standards and the function is preferably performed digitally in the IF or baseband. This demands wideband and high-performance analog-to-digital converters (A/D) for the uplink and digital-to-analog converters (D/A) for the downlink. In effect, the software radio base station extends the evolution of programmable hardware, increasing flexibility via increased programmability.

The baseband section is the most complicated part of a base station equipment. The high volume of processing involved often requires complex hardware and software design solutions, thus resulting in high costs, inflexibility to change, and restrictions in performance. This is a serious problem because the base station is the most numerous and costly part of a network, and base stations also have a critical impact on the end-to-end system performance. There are four types of processors commonly used in the design of communications equipment: the application specific integrated circuit (ASIC), the field programmable gate array (FPGA), the digital signal processor (DSP), and the general purpose processor (GPP). The GPP offers the most flexibility but has the lowest performance, whereas the ASIC is the least flexible but has the highest performance. Although ASIC can be used to minimize power consumption and the form factor and to maximize performance, they are not suitable for reconfigurable applications due to their inflexibility. FPGAs provide the best programmable solution for high speed signal processing functions that are highly parallel or involve linear processing. DSPs provide the best programmable solution for functions that involve complex analysis or decision-making.

Therefore, in software radio base stations, the FPGA, DSP, and GPP are normally used for baseband processing and ASIC can only be used for some common specific functions such as the up- and downconverters.

Generally speaking, there are five key selection criteria that should be considered when choosing an ASIC, FPGA, DSP, or GPP [17].

- *Programmability*: This is the ability to reconfigure a device to perform the desired functions for all of the target air interface standards.
- *Level of integration*: This is the ability to integrate several functions into a single device, thus reducing the size and hardware complexity of the digital radio subsystem.
- *Development cycle*: This is the time it takes to develop, implement, and test a digital radio function with a specific device.
- *Performance*: This is the ability of a device to perform a given function within the required time.
- *Power efficiency*: This is the power utilization efficiency of the device when performing the required function.

Each of these criteria has a direct impact on the decision a designer needs to make when designing a software radio. Table 2.1 gives an assessment of ASIC, FPGA, and DSP according to the above criteria [17]. It is seen that, with the current state of art, DSP offers the best compromise if one needs to choose only one type of components for software radio.

Table 2.1

Rating of Different Components for Software Radio

Evaluation Category	*ASIC*	*FPGA*	*DSP*
Programmability	1	4	5
Integration	2	5	5
Development cycle	5	1	3
Performance	5	4	3
Power efficiency	5	2	3

Since DSPs and GPPs have the highest programmability, the ultimate approach to a software radio base station could be to implement all the functionalities in the software and use DSPs and GPPs as the hardware platform. As an example, Figure 2.6 shows a software radio base station for WCDMA that was presented by PA Consulting, a British company based in Cambridgeshire, in 2003 [18]. PA Consulting's software radio base station employed an array of TigerScharc DSPs from Analog Devices to perform the baseband processing for WCDMA, such as channel estimation, path searching, spreading and despreading, the detection of the random access channel (RACH), and various symbol rate processing. For the coordination of the DSP array and the routing of data, PA Consulting adopted a GPP approach. As a software-only solution, this undoubtedly has the maximum flexibility and is particularly suited for micro and pico base stations due to its cost implications.

2.4.2 Software Architecture

The software architecture employed in software radios varies according to the hardware platform. Generally speaking, most vendors tend to separate the signal processing part from the control part due to their different nature.

Figure 2.6 PA Consulting's software radio-based WCDMA base station.

Figure 2.7 shows the software architecture used by PA Consulting [18]. The main DSP software runs on the DSP array and implements the primary functions in the baseband. It includes a block of core algorithms for the baseband to deal with special functionalities such as modulation and demodulation and coding and decoding. To support the DSP software, an algorithm library is employed to provide standard signal processing algorithms. The so-called Layer 1 software runs on the GPP, controls the DSP array, routes data traffic, and interfaces with the remainder of the base station. The common services act as the middleware between the DSP software and the real-time operating system (RTOS).

A similar software architecture used by Vanu Inc. for their software radios is shown in Figure 2.8 [19]. It is seen that two distinct parts, the control part and the signal processing part, are built upon a common operating system because a common hardware platform is used. The control part configures and controls the system and implements higher-level functions such as protocol state machines and network routing. The signal processing part implements the transforms between user data and a sampled representation of a RF waveform. The real-time operating system (RTOS) layer sits on top of the hardware platform. The advantage of employing a common operating system is that it isolates the signal processing application from the hardware and thereby significantly improves its portability.

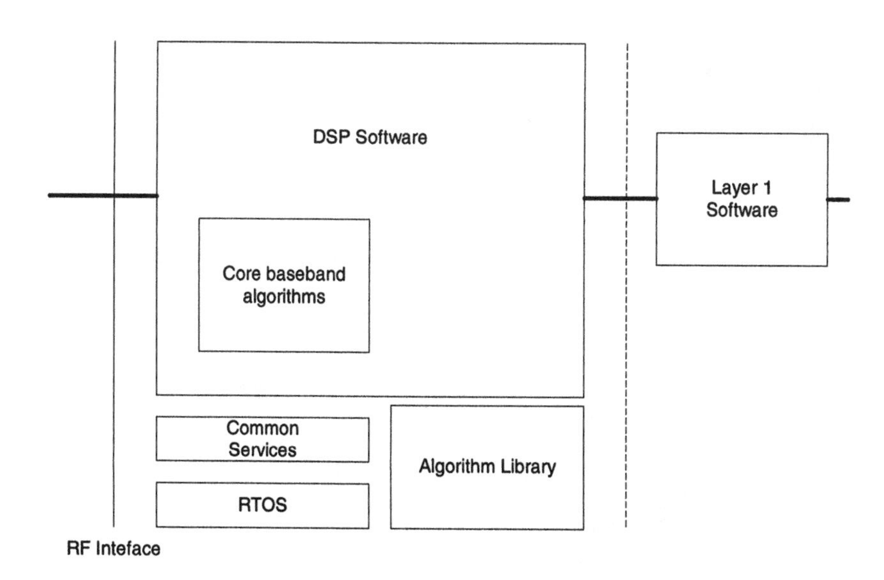

Figure 2.7 The software architecture for software radios used by PA Consulting.

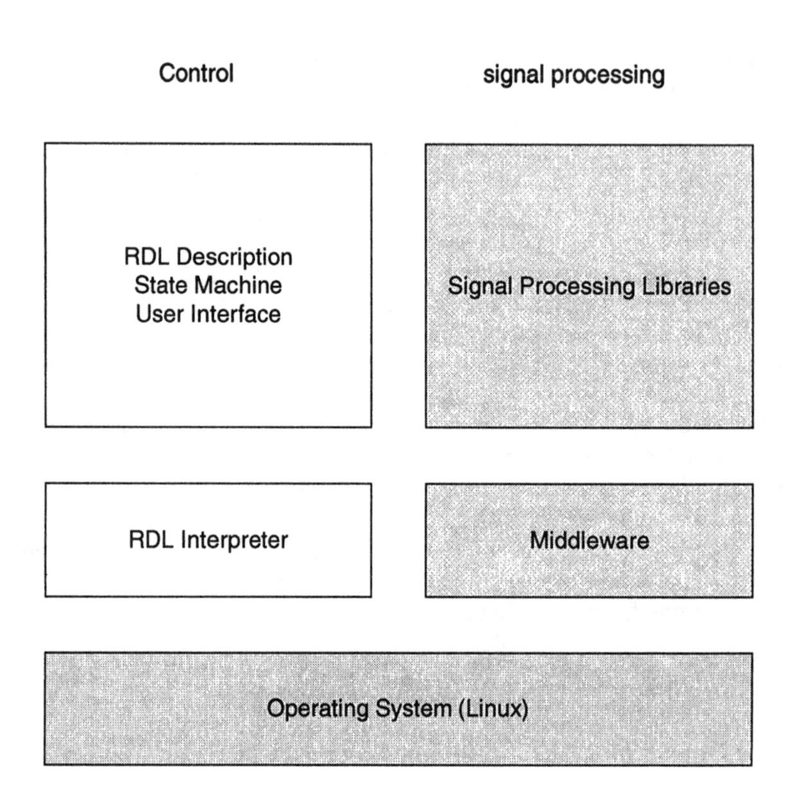

Figure 2.8 Software architecture of Vanu Inc.'s software radio.

In the Vanu system, the signal processing part consists of the middleware layer which performs data movement and module integration, and a library of signal processing modules for functions such as modulation, Viterbi encoding, and filtering. The control part consists of two blocks. The bottom block interprets state machines and processing descriptions, and the top block is a downloadable application that determines which waveform or air interface standard the system will implement. The downloadable application is written in a Vanu-developed language called RDL, the Radio Description Language.

Currently, there are still some tough challenges facing the software radio technology. To begin with, wideband antennas are needed to access multiple RF bands dynamically either sequentially or in parallel. The physical size of an antenna is closed related to the theoretical limit of its bandwidth and most antennas can only support bandwidths in the order of 10% or less. With the conventional antenna used in current radio access systems, it would be very difficult to support different systems whose carrier frequencies range from 800 MHz to 2.5 GHz. Admittedly, owing to the relatively relaxed restriction on antenna dimensions, this problem is less severe for base stations than for terminals. Next, the RF filters should be electronically tunable and offer great performance. Third, the performance of ADCs and DACs themselves must be improved further as it has strong impact on the behavior of the software radio in terms of scalability, signal-to-noise ratio, and power consumption. The fourth challenge is the speed of digital signal processors. The state-of-the-art DSPs are still not adequate to meet the demand of multistandard software radio base stations. Currently, many companies in the software radio community are working on novel multithread processors and process management strategies.

It should be pointed out that the analog-to-digital converters (ADC) are considered as particularly critical components for software radio base stations. The ADC takes the intermediate frequency (IF) signal, samples it at a rate of at least twice the frequency bandwidth, and translates each sample into a series of digital numbers. The following are key ADC characteristics [20]:

- *Sampling rate.* The sampling rate limits the bandwidth of the wideband analog RF. The bandwidth of the signal must be smaller than half the maximum sampling rate offered by the ADC because the Nyquist principle requires that the sampling rate be at least twice the frequency bandwidth.
- *Resolution.* The resolution of each sample depends on the number of bits used.
- *Spurious-free dynamic range (SFDR).* Spurious-free dynamic range is defined as the range of signal amplitudes that can simultaneously be processed without distortion or be resolved by a receiver without the emergence of spurious signals above the noise floor.

For the current software radio base stations, the ADC commands a sampling rate of at least 60 MHz, 14 bits, and a spurious-free dynamic range of more than 90 dB. Pushed by the demands from software radio design companies, ADC manufacturers are introducing products that need to meet those requirements. In the future, it is expected that ADC manufacturers will use superconducting technology that will operate at frequencies of several tens of gigahertz with SFDR between 120 and 160 dB. These ADCs will be able to translate microwave frequencies directly from the antenna into a digital bitstream. The high speed will enable the design of broadband software radio receivers that will cover bandwidths greater than 100 MHz. The high sensitivity could even remove amplifiers from the receiver chain if superconducting technology is employed [21].

2.5 CONCLUDING REMARKS

In this chapter, four types of emerging radio technologies for base stations have been addressed. Linearized transmitters use adaptive predistortion techniques to increase the efficiency of power amplifiers while maintaining linearity. Since the efficiency of the state-of-the-art power amplifiers is only around 14%, this technology is being employed and improved step by step by many power amplifier vendors. It is expected that the next milestone will be to increase the efficiency of power amplifiers to around 20% and then to around 40%. The superconducting filters and the cryogenic receiver front end use superconductors and cryogenic techniques to increase the performance of the preselect filter and to lower the thermal noise level of the receiver. This will help reduce mobile terminal transmit power and increase cell capacity and coverage. Another advantage of superconducting filters and the cryogenic receiver front end is that they can be installed as add-on appliqué modules without changing the configuration of the existing system. The advantage of the superconducting filters can be further exploited to enable dynamic spectrum allocation or more efficient spectrum utilization in regions where the spectrum is not tightly regulated, such as the United States. The remote RF head and radio over fiber offer a lot of benefits to mobile network operators. It is expected that the recently published industrial specification on common public radio interface (CPRI) will facilitate the take-up of such products. The interest in software radio has been mainly from mobile terminal vendors due to the fact that mobile terminals need to roam between different systems frequently. However, there are significant benefits in employing soft radio technology in base stations, and several vendors are working on the technology vigorously. To a large extent, the future of such products depends on the progress of digital signal processors, ADCs, and DACs, as well as wideband RF devices. Owing to practical factors such as the balance of long-term and near-term cost, and the performance of current RF and digital components, however, software radio should be regarded as an evolving technology trend instead of a revolution. In fact, many ideas of software radio have been implemented in commercial base stations for the third generation systems, including WCDMA. The software radio technology will help minimize network operators' cost of future network upgrades, from introducing new carriers to migrating to HSDPA.

In fact, these four technologies are not independent. Instead, they can be integrated together to form advanced radio access systems in the future. For instance, the radio heads can be equipped with linearized transmitters and cryogenic receivers, both having wide bandwidth. The base station server can be built on software radio hardware and software architecture, and be connected to the radio head via an optical fiber network. Such a system will be future-proof, have great and flexible coverage without using excess power, and result in lower operation and maintenance expenditures for network operators.

References

[1] L. Hanzo, W. Webb, and T. Kelly, *Single- and Multi-Carrier Quadrature Amplitude Modulation*, New York: John Wiley & Sons, 2000.

[2] P. B. Kenington, *High-Linearity RF Amplifier Design*, Norwood, MA: Artech House, 2000.

[3] P. B. Kenington, "Linearized Transmitters: An Enabling Technology for Software Defined Radio," *IEEE Communications Magazine*, Vol. 40, No. 2, February 2002, pp. 156 – 162.

[4] http://www.suptech.com.

[5] http://www.conductus.com.

[6] H. Li, et al., "A Demonstration HTS Base Station Sub-System for Mobile Ccommunications," *Superconductor Science and Technology*, Vol. 15, 2002, pp. 276 – 279.

[7] J. Mazierska and M. V. Jacob, "High Temperature Superconducting Filters for Mobile Communications," *Proceedings of the International Symposium on Recent Advances in Microwave Technology, ISRAMT'99*, Malaga, December 1999.

[8] http://pr.fujitsu.com/jp/news/2002/09/20.html.

[9] http://www.iscointl.com.

[10] www.teledotcom.com/article.

[11] K. Watanabe, et al., "Base Station for WCDMA – System Design and Product Series," *NEC Research & Development*, Vol. 42, No. 4, Oct. 2001, pp. 354-359.

[12] A. G. Gonzalez, P. R. Aragon, and M. G. Roa, "Radio on Optical Fiber Repeaters for Cellular Mobile Communications," http://www.tid.es.

[13] D. Wake, "Trends and Prospects for Radio over Fiber Picocells," http://www.osda.org.uk/downlods/mwp02_DWake.pdf.

[14] Common Public Radio Interface, http://www.cpri.info.

[15] J. Mitola III, *Software Radio Architecture: Object-Oriented Approaches to Wireless Systems Engineering*, New York: John Wiley & Sons, 2000.

[16] P. G. Cook, "Network Oriented Base Stations," *SDR Forum Document*, SDR-01-I-0057-V0.00, July 2001.

[17] L. Pucker, "Paving Paths to Software Radio Design," *CommsDesign*, January 2004.

[18] A. Carr, "The All Software 3G Basestation," white paper, PA Consulting.

[19] Vanu Software Radio Basestation, http://www.vanu.com.

[20] L. Luneau and F. Luneau, "A Software-Defined Radio Architecture for Wireless Hubs," white paper, Radical Horizon, February, 2002.

[21] A. Fujimaki et al., "Broadband Software-Defined Radio Receivers Based on Superconductive Devices," Department of Quantum Engineering, Nagoya University, Japan.

Chapter 3

Mobile Terminal Positioning

It is expected that location services (LCS), or location-based services (LBS), will bring great convenience and new exciting services to subscribers of future mobile communications networks and therefore generate significant revenues to the operators. LCS requires the integration of wireless network infrastructure, mobile terminals, and a range of location-specific applications and content. The fundamental technology supporting LCS, however, is mobile terminal positioning. The GSM location services standards initiative was originally triggered by the Federal Communications Commission (FCC) mandate requiring U.S. wireless operators to ensure that callers requesting public emergency services can be located reliably, which is called wireless E911. Since location finding is a stochastic process, it is very difficult to achieve guaranteed high accuracy. Therefore, the statistical requirement shown in Table 3.1 was specified for E911 services by FCC in 1999 [1].

Table 3.1

Accuracy Required for E911 Services

Solutions	*67% of Calls*	*95% of Calls*
Terminal-based	50 meters	150 meters
Network-based	100 meters	300 meters

The 3G standardization activities of the late 1990s predicted some new location services for the public, in addition to emergency services. These new location services were direction finding, localized advertising, and clubbing. Furthermore, it was envisaged that corporate customers would demand such location services as tracking and workforce and fleet management. To some extent, this promoted the technology development of location services in UTRAN.

To enable location services, the position of the mobile terminal in question must be found first. Owing to the complexity of mobile environment, positioning a mobile terminal accurately is a very challenging task. Despite a decade of effort, the mobile communications community is still searching for cost-effective and reliable techniques for mobile terminal positioning. In this chapter, an overview on major mobile terminal positioning techniques is given first. Then the implementation issues and the advantages and disadvantages of different positing techniques specified in UTRAN are discussed. Finally, the UTRAN LCS architecture and operating principles are presented.

3.1 OVERVIEW OF POSITIONING TECHNIQUES

There are five main stream mobile terminal positioning techniques, namely, (1) cell ID, (2) angle of arrival measurement, (3) time of arrival measurement, (4) time difference of arrival measurement, and (5) assisted GPS [2–4]. In the following, the principles of these methods are described and the pros and cons of each method are discussed.

3.1.1 Cell ID

The simplest method to position a mobile terminal is to use the cell ID that is specific to the cell coverage area where the mobile terminal is located. This information can be obtained by paging, call update, and network registration area update. When the mobile terminal is in the state of soft handover in a CDMA network, it is normally associated with several cell areas. In this circumstance, the radio network controller (RNC) needs to decide on the most appropriate cell ID based on such information as handover history and signal strength. A major disadvantage of the cell ID method, however, is that the position error can be as large as the cell size. For instance, a picocell could be 150m in radius and a Macro cell could be 20,000m in radius. Clearly, errors of such magnitude are not always acceptable and, with the cell ID method, it is impossible to achieve the 100m accuracy reliably.

To improve the accuracy of cell ID method, some additional means can be employed, which include measuring the signal round trip delay (RTT), the signal strength, the signal characteristic pattern, and multipath characteristics of the radio signals arriving at a cell site from the terminal. For measuring the signal strength, it is beneficial to use signals from multiple cell sites to mitigate the multipath effects. For measuring the signal characteristic pattern, pattern recognition techniques can be used to identify the unique radio frequency pattern or "signature" associated with the location. Compared with other techniques, RTT measurement appears to be the simplest and the most reliable. In UTRAN, RTT is defined as the time difference between the transmission of the beginning of a downlink dedicated physical channel (DPCH) frame to a user equipment (UE) and the reception of the beginning of the corresponding uplink DPCH from the UE. Obviously, the RTT observed by a base station includes the response delay that occurred at the mobile terminal. Therefore, it is mandatory for the UE operating in FDD mode to report RX-TX (receive-transmit) time difference, which is defined as the difference between the time when the UE begins uplink DPCH frame and the time when the downlink DPCH frame is first detected.

3.1.2 Angle of Arrival Measurement

Another mobile terminal positing method is to use the antenna array at the base station to measure the angle of arrival (AoA) of the incoming signal from the mobile terminal. The intersection of two directional lines of bearing measured at two different base stations gives the position of the mobile (see Figure 3.1). Naturally, this technique requires a minimum of two base stations for each location measurement. If possible, more than two base stations should be used to reduce the estimation error by employing techniques such as the least mean square (LMS) method.

Unfortunately, there are some practical problems associated with the angle of arrival measurement technique. First, the accuracy of direction measurement is directly proportional to the size of the antenna array involved, so direction finding needs antenna arrays of reasonable size at the base stations. Owing to reasons to be discussed in Chapter 5, antenna arrays are not likely to become common in cellular networks soon. Even when smart antenna arrays are introduced, the accuracy of AoA estimates may not meet the requirement for positioning. This means that the AoA method will not become widely available in the near future. Second, owing to the multipath phenomena, it is not always possible for the network to distinguish between the signals directly from the terminal and the ones reflected by surrounding buildings. Admittedly, the latter problem is a general one for positioning and is not specific to the AoA method.

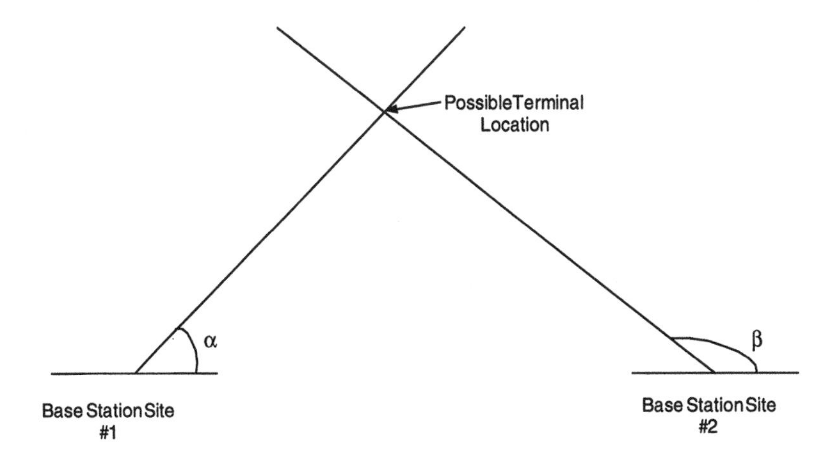

Figure 3.1 Positioning using AoA.

3.1.3 Time of Arrival Measurement

A further terminal positioning method is to exploit the time of arrival (ToA) information to locate the mobile position based on the intersection of the distance circles (see Figure 3.2). The propagation time of the radio wave is directly proportional to the distance that it travels, and multiplying the speed of light with the time gives the distance from the mobile terminal to the base station. Obviously, it requires three time measurements to determine a unique position by triangulation. This principle has been used by the global positioning system (GPS), where a circle becomes a sphere in space and an additional measurement is required due to the receiver-clock bias. The bias is caused by the unsynchronized clocks in the receiver and the satellite. Similarly, for terrestrial radio access networks, it is desirable to have synchronized or known clocks for all transmitters and receivers. Otherwise, a 1-μs timing error could lead to a 300-m position error. UTRAN networks, especially those in FDD mode, are unsynchronized. Therefore, the ToA method is only used in the techniques based on the global positioning system.

3.1.4 Time Difference of Arrival Measurement

To mitigate the problem of estimating the absolute time used for the radio wave to reach the mobile terminal from a base station, the time difference of arrival (TDOA) method can be employed, which determines the mobile terminal position based on the time difference measurement rather than the absolute time measurement. TDOA is often referred to as the hyperbolic system, as the time difference can be converted to a constant distance difference to two base stations (as foci) to define a hyperbolic curve. The intersection of two hyperbolas determines the position. This means that two pairs of base stations (at least three) are needed for positioning (see Figure 3.3). In UTRAN, the TDOA technique is referred to as the observed time difference of arrival (OTDOA). Its principles of operation will be explained in Section 3.2.2 and the mathematical details will be given in Appendix 3A. It should be pointed out that the accuracy of all the radio based techniques discussed above is affected by cochannel interference, blockage (no line of sight), and multipath that are common in mobile communications environment. It is a great challenge to mitigate these adverse effects. The solutions often lie in the combination of different complementary techniques together with sophisticated signal processing.

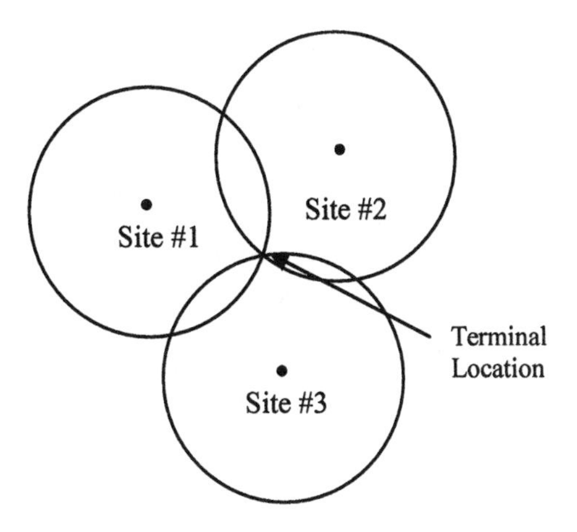

Figure 3.2 Positioning using ToA.

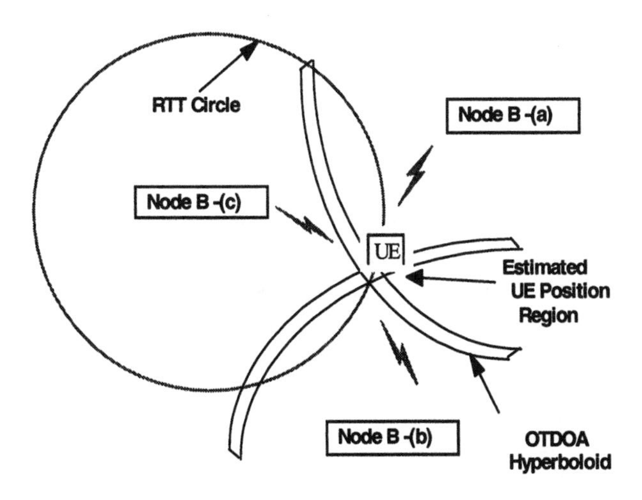

Figure 3.3 Positioning using TDOA.

3.1.5 Assisted GPS

As the performance of the satellite-based GPS receiver is improving rapidly and the receiver size and price keep decreasing, it becomes convenient and economical to implement an assisted GPS (A-GPS) solution for mobile terminal positioning, which requires software and hardware modifications to both the mobile and the network. GPS provides an affordable means to determine position, velocity, and time around the globe. The satellite constellation is developed and maintained by the U.S. Department of Defense. Civilian access is guaranteed through an agreement with the Department of Transportation. GPS satellites transmit two carrier frequencies. Typically, only one is used by civilian receivers. From the perspective of these civilian receivers on the ground, GPS satellites transmit at 1575.42 MHz using the code division multiple access (CDMA) technique, which uses a direct sequence spread spectrum (DS-SS) signal at 1.023 Mcps with a code period of 1 ms. Each satellite's DS-SS signal is modulated by a 50-bps navigation message that includes accurate time and coefficients (ephemeris) to an equation that describes the satellite's position as a function of time. The receiver position determination is based on the time of arrival (TOA).

The functions of the GPS receivers include the following:

- Measuring the distance from the satellites to the receiver by determining the pseudo ranges (code phases);
- Extracting the time of arrival of the signal from the contents of the satellite transmitted message;
- Computing the position of the satellites by evaluating the ephemeris data at the indicated time of arrival;
- Calculating the position of the receiving antenna and the clock bias of the receiver by using the above data.

Position accuracy at the receiver is determined by the satellite clock, satellite orbit, ephemeris prediction, ionospheric delay, and tropospheric delay. To reduce estimation errors, range and range-rate corrections can be applied to the raw pseudorange measurements in order to create a position solution that is accurate to a few meters in open environments. The most important correction technique is differential GPS (DGPS). It uses a reference receiver at a surveyed position to send correcting information to a mobile receiver over a communications link. Generally speaking, it is difficult to employ a traditional autonomous GPS receiver in mobile terminals for the following reasons. First, its start-up time, which is defined as the period from turning on to the initial position fix, is relatively long due to the long acquisition time of the navigation message ranging from 30 seconds to a few minutes. Second, the GPS receiver is incapable of detecting weak signals that result from indoor and urban canyon operations. Third, the power consumption of the autonomous GPS receiver is relatively high per fix, primarily due to the long signal acquisition time. To deal with these problems, the assisted GPS method shown in Figure 3.4 can be employed.

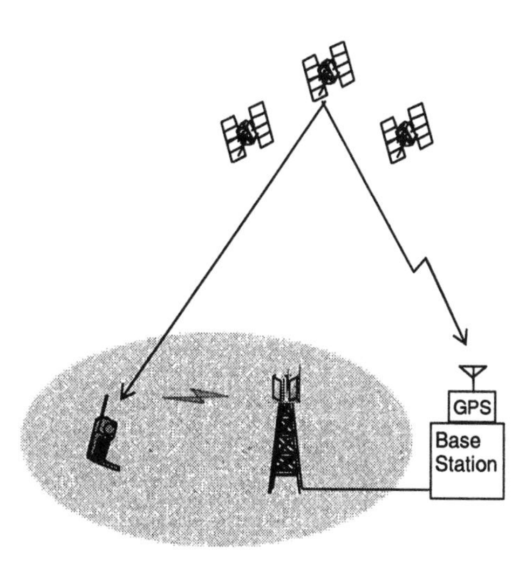

Figure 3.4 Assisted GPS positioning.

The basic idea of the assisted GPS is to establish a GPS reference network (or a wide area DGPS network) whose receivers have clear views of the sky and can operate continuously. This reference network is also connected to the cellular infrastructure, and it continuously monitors the real-time constellation status and provides precise data such as satellite visibility, ephemeris and clock correction, Doppler, and even the pseudorandom noise code phase for each satellite at a particular epoch time. Upon the request of the mobile terminal or location-based applications, the assist data derived from the GPS reference network are transmitted to the GPS receiver in the mobile terminal to aid fast start-up and to reduce terminal power consumption. The reduction in acquisition time and power consumption is due to that fact that the Doppler versus code phase uncertainty space is much smaller than that in a conventional GPS receiver as a limited search space has been predicted by the reference receiver and the network. This allows for rapid search speed and for a much narrower signal search bandwidth, which enhances sensitivity. Once the embedded GPS receiver acquires the available satellite signals, the pseudorange measurements can be delivered to network-based position determination entity (PDE) for position calculation or can be used directly by the terminal to compute its position.

Additional assisted data, such as DGPS corrections, approximate terminal location, or cell base station location, and other information such as the satellite almanac, ionospheric delay, and universal time coordinated (UTC) offset, can be transmitted to improve the location accuracy, decrease acquisition time, and allow for terminal-based position computation. To reduce the number of bits necessary to be exchanged between the terminal and the network, one can choose to transmit only the parameter changes instead of the raw parameters themselves.

The disadvantage of the A-GPS is that, in addition to the requirement of a GPS reference network and additional location determination units in the network, the mobile terminal must be equipped with, at a minimum, a GPS antenna and RF downconverter circuits. Also, it must make provisions for some form of digital signal processing software or dedicated hardware. These inevitably increase the terminal cost.

3.2 POSITIONING TECHNIQUES IN UTRAN

Three positioning techniques have been standardized in UTRAN: Cell ID, network-assisted GPS positioning, and OTDOA with optional network assistance using idle periods. In addition, the uplink TDOA (U-TDOA) has been included in the GERAN specification. Since the cell ID-based technique is rather straightforward, as explained in Section 3.1, discussions are focused on the assisted GPS, OTDOA, and U-TDOA in the following sections.

3.2.1 Assisted GPS

There are two types of A-GPS techniques specified in the UTRAN standards: the network-based UE-assisted GPS solution and the network-assisted mobile terminal-based GPS solution. In the former, the calculation of the mobile position is carried out in the network, whereas in the latter the calculation is done at the mobile terminal first and the result is sent to the network. Despite the above classification of two assisted GPS solutions, their principles are the same. If the GPS receiver does not know its approximate location, it will not be able to determine the visible satellites or estimate the range and Doppler frequency of these satellites. It has to search the entire code phase and frequency spaces to locate the visible satellites. For the code phase space, it spans from 0 to 1,023 chips. For the frequency space, it spans from –4.2 kHz to +4.2 kHz. The relative movements between the satellites and receiver make the search even more time-consuming. Therefore, the time-to-first-fix (TTFF) is one important parameter to evaluate the quality of a GPS receiver. For autonomous GPS, the present state-of-the-art TTFF for an un-initialized GPS receiver is approximately 60 seconds. Clearly, this is unacceptable for certain applications such as E911 and it would make the mobile terminal too power consuming. By transmitting assistance data over the mobile communications network, the TTFF of a receiver can be reduced to a few seconds. This is achieved by significantly reducing the search window of the code phase and the frequency space by sending precise measurements of the common parameters to the user equipment (UE) or terminal from the network.

The reduction in search space allows the receiver to spend its search time focusing on where the signal is expected to be.

3.2.1.1 UE-Assisted GPS

The network-based UE-assisted solution shifts the majority of the signal processing in the traditional GPS receiver to the network processor. This method requires an antenna, RF circuits, and some digital processing capability in the mobile terminal for making measurements by generating replica codes and correlating them with the received GPS signals. The network transmits a short assistance message to the mobile terminal, consisting of time, visible satellite list, satellite signal Doppler, and code phase search window. These parameters help the GPS receiver embedded in the mobile terminal reduce the acquisition time. These assistance data are valid for a few minutes. The pseudorange data created by the GPS receiver is sent from the UE. Upon receiving the pseudorange data, the corresponding network processor or location server estimates the position of the UE. The differential correction (DGPS) can also be applied to the pseudorange data available to the network to improve the position accuracy.

3.2.1.2 UE-Based GPS

The network-assisted UE-based solution maintains a fully functional GPS receiver in the terminal. This requires the same functionality as described in terminal-assisted GPS, plus additional means for computing the position of the mobile terminal. This additional terminal function generally adds to the terminal's total memory (RAM, ROM) requirements in addition to the MIPS. In the initial start-up scenario, more precise information about the satellite orbital elements (ephemeris) must be provided to the UE than for the network-based UE-assisted case. In the case of ephemeris data being transmitted to the terminal, this data is valid for 2 to 4 hours or more and can be updated as necessary over time. Therefore, once the terminal has the data, subsequent updates are infrequent. Apart from point-to-point transmission, a broadcast channel can be used to distribute this data efficiently to all terminals in a network. If better position accuracy is required for certain applications, differential correction (DGPS) data must be transmitted to the terminal approximately every 30 seconds. The final position of the mobile terminal is generated inside the terminal itself. The calculated UE location can then be sent to an application outside of the mobile terminal if required.

3.2.2 OTDOA

In UTRAN, it is mandatory to employ the basic method of observed time difference of arrival (OTDOA) using super frame number-super frame number (SFN-SFN) measurements.

OTDOA is basically a time difference measurement technique (TDOA) to locate a mobile by triangulation. Each OTDOA measurement for a pair of downlink transmissions describes a line of constant difference along which the mobile terminal may be located. The terminal's position is determined by the intersection of these lines for at least two pairs of base stations (node Bs) (see Figure 3.3). The accuracy of the position estimates achieved with this technique depends on the precision of the timing measurements, the relative position of the node Bs involved, and is also subject to the effects of multipath radio propagation. The best results are normally obtained when the distances between the mobile terminal and the base stations are similar. If this is not the case, the accuracy will be reduced, which is sometimes referred to as the geometric dilution of position (GDP).

In order to estimate the UE position, the calculation function needs to know the following: (1) OTDOA measurements; (2) surveyed geographic positions of the node Bs whose signals are measured; and (3) actual relative time difference (RTD) between the transmissions of the node Bs at the time the OTDOA measurements were made. The accuracy of each of these measurements contributes to the overall accuracy of the position estimate.

There are several approaches to determining the RTD. One is to synchronize the transmissions of node B. In this technique, the RTD are known constant values that may be entered into the database and used by the calculation function when making a position estimate. The synchronization accuracy must be in the order of tens of nanoseconds, as 10-ns uncertainty contributes to 3-m error in the position estimate. Drift and jitter in the synchronization timing must also be well controlled because these also result in uncertainty in the position estimate. Synchronization to this level of accuracy can be achieved by installing GPS receivers at the Node Bs. In the UTRAN time division multiplexing (TDD) mode, the Node Bs are generally synchronized.

In the UTRAN frequency division duplex (FDD) mode, node Bs may be unsynchronized and left to run freely within some constraints. In this scenario, the RTD will change slowly with time. The rate of change depends on the frequency difference and jitter between Node Bs. If, for example, the maximum frequency difference between two Node Bs is $\pm 10^{-9}$, then the start of transmission of a 10-ms code sequence will drift through a cycle in about 1,390 hours (or 57 days). With this relatively slow rate of drift, the measurements required to obtain the RTD can be performed by fixed location measurements units (LMU) at known positions and stored in the database for use by the calculation function. Alternatively, the location measurement units required to support RTD estimation may be colocated with the Node Bs. In UTRAN, the primary OTDOA measurements made by the UE are sent via signaling over the Uu, Iub, and Iur interfaces between the UE and the serving RNC. The calculation function makes use of the measurements, the known positions of the transmitter sites, and the RTD of the transmissions to estimate the UE's position.

Since the terminal positioning involves several difficult measurements, there is always some uncertainty in the results. Physical conditions, errors, and resolution limits in the apparatus all contribute to the uncertainty. To minimize the uncertainty in the UE positioning result, it is important that the UE is provided as many measurements of OTDOA as possible. Then the UE can provide SFN-SFN observed time difference measurements for as many cells as it can receive signals from. The cells to be measured should include those in the active set and the monitored set for handover. In order to support the OTDOA method, the positions of the UTRAN transmitters (base station antennas) need to be accurately known by the calculation function in SRNC or stand-alone SMLC (SAS) or UE. This information may be measured by appropriate conventional surveying techniques.

The location estimate is performed by a position calculation function (PCF) located in the network or in the UE itself. With the same network architecture, UE functions, LMU functions and measurement inputs, the PCF can be based on one of two possible variants of OTDOA, known as "circular" and "hyperbolic" variants. For interested readers, the technical details of the circular variant are given in Appendix 3A.

3.2.3 Hearability Problem and Countermeasures

The OTDOA technique relies on the fact that terminals can receive signals from a number of geographically dispersed base stations. Unfortunately, this may not always be the case in practice. Transmitters in a CDMA network share a common frequency channel all the time, and closer transmitters tend to "drown out" more distant ones, making it impossible for the mobile terminal to measure the signals from those more distant transmitters. In other words, a terminal close to its serving Node B may not hear other Node Bs on the same frequency, which is termed the "hearability" problem in UTRAN [5]. Up until now, two solutions to the hearability problem have been proposed to the UTRAN standardization body 3GPP: the idle period in the downlink (IPDL) [6, 7] and the cumulative virtual blanking (CVB) [5, 8]. The former has been standardized in UTRAN, whereas the latter is being considered as a candidate for future standards.

3.2.3.1 IPDL

In IPDL, each node B is configured appropriately by the controlling RNC (CRNC) using the NBAP (node B application part) protocol to cease its transmission for short periods of time called idle periods. During an idle period, terminals within a cell are free from the interference caused by the serving node B so they can measure other node Bs with much higher accuracy. The UEs are made aware of the occurrences of IPDLs by using radio resource control (RRC) signaling over the Uu interface. Also, during idle periods, the real-time difference measurements can be carried out. Because the IPDL method is based on the downlink, it can be exploited by a large number of mobile terminals simultaneously.

The idle periods are arranged in a predetermined pseudorandom fashion according to some parameters. In general there are two modes for the idle periods: continuous mode and burst mode. In the continuous mode, the idle periods are active all the time. In burst mode, however, the idle periods are arranged in bursts where each burst contains enough idle periods to allow a UE to make sufficient measurements for its location to be calculated.

3.2.3.2 CVB

Allocating idle periods in the downlink has one major problem: All the downlink traffic from the Node B is blocked during the idle periods so the network suffers from downlink capacity loss. To avoid this problem, an alternative method known as cumulative virtual blanking (CVB) has been proposed to 3GPP by a British firm, Cambridge Positioning System (CPS) [5, 8]. Instead of physically turning off the Node B transmissions for short idle or blanking periods, the CVB method employs some means of digital signal processing in the serving mobile location center (SMLC) to remove stronger signals successively and then extract the timing information for weaker signals. In principle, the CVB method may achieve results similar to IPDL without the capacity loss.

The CVB method is based on different measurements from those used in conventional OTDOA; each node B and UE need to capture a "snapshot" of the downlink signal when a location calculation is required. To implement the CVB technique, the SMLC requires the following measurements to be made: (1) a snapshot of the radio channel received by the UE; and (2) the time coincident snapshots of the transmitted radio channel of each node B. Each snapshot is typically 1 kB at the UE and 4 kB at the node B, and it represents the radio channel baseband I-Q signals. There is no need for the UE to process the captured snapshot. The snapshot is simply buffered and transferred to the SMLC via the Uu interface. It is not necessary for the UE to calculate SFN-SFN timings nor to have any knowledge of the neighboring node Bs. The SMLC preprocesses the coincident snapshots from the UE and each relevant node B by successively correlating the UE snapshot with the node B snapshots and removing the strongest node B signal from the UE signal by subtraction. The output of this preprocessing stage is a set of OTDOA measurements, which are transferred to the standard positioning algorithms instead of the SFN-SFN measurements that would have been obtained using the standard OTDOA measurement procedures.

Figure 3.5 shows the basic block diagram illustrating how the CVB function might be integrated into a UE. The modules that are potentially affected could be the analog-to-digital converter (A/D) and the buffer, as higher performance A/D and larger buffer size may be needed. Given the higher performance of UTRAN terminals, however, it is envisaged that the addition of the CVB functions in the UE will probably require only software modifications.

Figure 3.6 shows a basic block diagram illustrating how the LMU function supporting CVB can be incorporated into a Node B. The key elements are the sampling device, the snapshot buffer, the time source, a method of accurately time stamping the snapshot measurement, and a method of roughly synchronizing the sampler between Node Bs. In this configuration, the downlink radio channel is sampled before being upconverted and transmitted. In order that there is a complete overlap between the UE snapshot and the node B snapshots, the node B snapshot needs to be long enough to take account of measurement timing uncertainties and Node B synchronization errors. The time stamp is an important element in the Node B LMU function. It is necessary to time stamp the data snapshot accurately as the time stamp errors translate directly into position errors. There are several ways to achieve accurate time stamping, the most obvious being the use of GPS time.

The operation procedures of CVB can be described as follows:

- A position request from the core network (CN) is routed to the SRNC, which initiates an RRC measurement request to the UE. The request includes timing information instructing the UE when to capture the snapshot.
- The UE captures the data and returns it in an RRC measurement response.
- The SRNC passes the data to the SMLC.

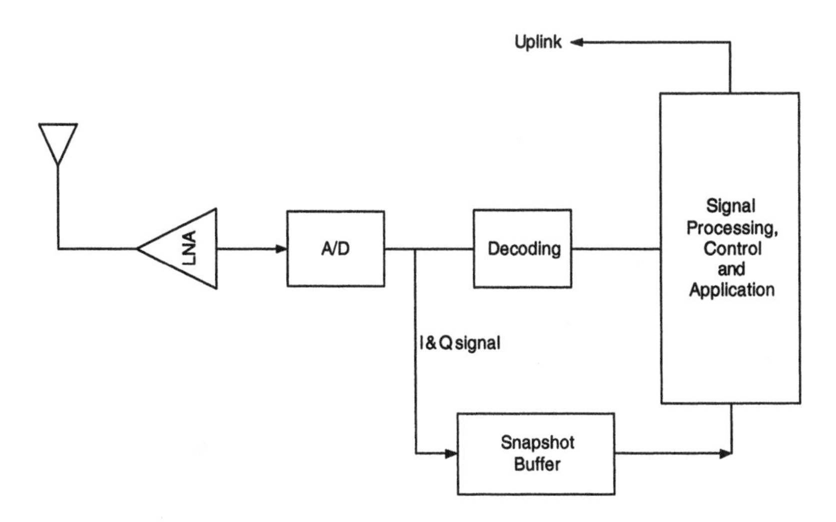

Figure 3.5 UE block diagram with CVB.

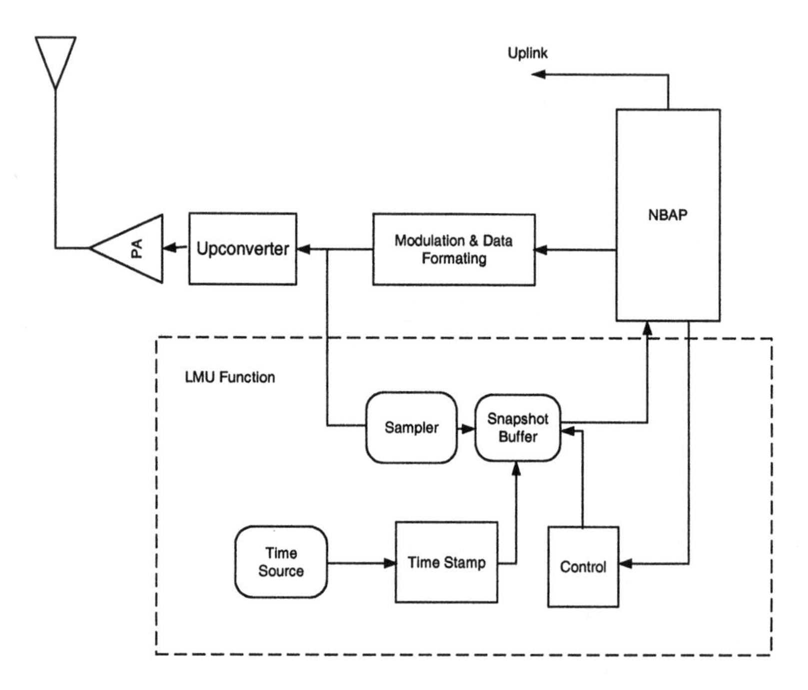

Figure 3.6 A node B with integrated CVB functions.

- The SMLC cross-correlates the snapshot made at the UE with the snapshots from the relevant node Bs. The time offset for the strongest node B correlation is measured.
- An estimate of the signal received by the UE from the strongest node B is made and subtracted from the UE snapshot.
- Step 4 is carried out for the second strongest node B. This is iterated until no further signal can be extracted.

Then, the time offsets gathered are transferred to the standard OTDOA location procedure, which calculates the UE position and responds to the original measurement request with the result.

The primary advantages of OTDOA with CVB compared to OTDOA with IPDL are the avoidance of the network capacity loss and the reduced implementation complexity. These advantages arise from the use of the cumulative virtual blanking in the digital signal processing (DSP) domain rather than the use of physical blanking at the transmitter. Physical disabling of the transmitter completely suppresses the interfering signal, whereas virtual blanking attenuates the interfering signals rather than actually removing them entirely. Some tests carried out by CPS indicated that attenuation of up to 40 dB is achievable. When a UE is very close to a particular Node B, it may happen that the relatively strong received signal from that site will be so dominant that even 40 dB of attenuation is insufficient to overcome the near-far problem. This is likely to occur when the UE is very close to the Node B in a macro-cell scenario, but such a scenario actually defines a relatively small geographical range which might be good enough for the location services in question.

3.2.4 Uplink TDOA

Another positioning technique, uplink TDOA (U-TDOA) proposed by a U.S. firm, TruePosition, has been adopted for GERAN by 3GPP [9, 10]. The difference between the U-TDOA and the OTDOA is that in the former the measurement is done by the network using the uplink signals from the UE to the base stations, whereas in the latter the measurement is done at the UE using signals from the base stations to the UE. The advantages of U-TDOA are the following. First, no IPDL is needed, so the downlink capacity is preserved. Second, since the signal processing is done in the network equipment, more complicated processing becomes affordable and therefore the positioning accuracy can be potentially improved. Third, there is no need to modify the UE nor the node B. Fourth, the impact on the standards is low. Fifth, it provides some protection against UE obsolescence and eases the network management, as it is far easier and less complex to upgrade the software in RNCs and LMUs than it is to upgrade millions of terminals.

Owing to the cost and power consumption limit, mobile terminals have a limited amount of processing power for location finding. Network-based location systems can combine the DSP power from many LMUs for the location of a single mobile station and, as a result, are capable of doing much more processing over much longer periods of time.

For mobile terminals in active transmission, no additional signal is needed for location determination. The Uplink TDOA technique locates the UE using the energy associated with the existing bearer and control at the normal power level. No power-up at the terminal is needed due to the long integration time. For mobile terminals in the idle mode, the transmission needed for U-TDOA is no more than the normal subscriber traffic and for a much shorter period of time (100–500 ms). In fact, when accessing the network, the mobile terminal normally uses higher power. Therefore, it is preferable to use the access channel for U-TDOA measurements.

It should be pointed out that there are serious challenges in using Uplink TDOA location systems in CDMA networks when the mobile terminal is in close proximity to the serving Node B. In this scenario, it will be very difficult for other node Bs or LMUs to measure the uplink signal from the terminal, which is the classical "near-far" problem. Undoubtedly, there are techniques that can mitigate the near-far problem in CDMA networks, one of which is the so-called interference canceler. When dealing with each desired user, the interference canceler treats all other signals in the cell, or any known signals in the adjacent cells, as interference and attempts to remove them before decoding the desired signal. With the current state of art, this technique appears to be uneconomical and too complex to be justified for the operation of U-TDOA.

The solution proposed by TruePosition to solve the "near-far" problem is to employ the processing gain inherent in CDMA systems. They propose to use integration time up to 10,000 user data bits. At a spreading ratio of 256, an integration over 2.5 million chips will be computed, compared to the 256 – 1,024 chip integration as suggested for OTDOA. This long integration time, made possible by the powerful DSPs in the LMUs, should allow any node B site, which is contributing to the positioning process, to detect this signal at a level that is 40 dB below the level received by the serving node B. Assuming a suburban propagation law, where the signal attenuation is proportional to $1/r^{3.5}$, a near/far ratio of 14 is achieved. Therefore, as long as the UE is more than 1/14 of the distance of the cell site spacing from the serving node B, the surrounding node Bs should be able to detect the UE.

3.3 UTRAN LCS ARCHITECTURE

To facilitate the adoption of LCS, various functional entities, signaling messages, and procedures have been specified in UTRAN by 3GPP [11, 12]. This section presents a high-level overview on these issues and shows how the LCS technology is implemented in UTRAN.

Figure 3.7 illustrates the general architecture for UE positioning in UTRAN. The key entities involved in the positioning process include the UE, the location measurement unit (LMU), and the serving mobile location center (SMLC). When the SMLC is implemented as a detached entity, it is referred to as stand-alone SMLC (SAS). Communications among the UTRAN UE positioning entities make use of the messaging and signaling capabilities of the UTRAN interfaces (Iub, Iur, and Iupc). The serving radio network controller (SRNC) receives authenticated requests for UE positioning information from the core network (CN) across the Iu interface. SRNC manages the UTRAN resources including node Bs, location measurement units, the SMLC, the UE, and the position calculation functions in order to estimate the location of the UE. SRNC may also make use of the UE positioning function for internal purposes such as position-based handovers.

3.3.1 LCS Operations

A schematic functional description of LCS operations in UMTS is shown in Figure 3.8. Upon request from the core network, the UE positioning function in the radio network controller (RNC) takes the following actions:

- Requesting measurements, typically to the UE and one or more Node Bs;

- Sending the measurement results to the appropriate calculating function within UTRAN;
- Receiving the result from the calculation function in UTRAN;
- Performing any needed coordinate transformations;
- Sending the results to the LCS entities in the CN or to application entities within UTRAN.

In the event that the client is internal to UTRAN, the request may be made directly to the UTRAN UE positioning entities as the internal clients are considered to be "preauthorized."

As part of its operation, the UTRAN UE positioning calculation function may require additional information. This may be obtained by communicating directly with a database, or it may be through a request to UTRAN UE positioning entities that mediate the request and the return of information from the appropriate database.

It is possible that some independent information is available to UTRAN, which can be either the direct positioning information, or some auxiliary information to assist the calculation function. The UTRAN UE positioning co-ordination function, as part of its activity to supervise the positioning process, may query the UE or other elements of the UTRAN to determine their capabilities and use this information to select the mode of operation.

3.3.2 Location Measurement Unit

The location measurement unit (LMU) is responsible for radio measurements to support one or more positioning methods and it communicates the results to a RNC. These measurements fall into one of the two categories: (1) positioning measurements that are specific to one UE and used to calculate its position; and (2) assistance measurements that are applicable to all UEs in a certain geographic area.

They may also perform calculations associated with the measurements. All positioning and assistance measurements obtained by a location measurement unit are supplied to a particular control RNC (CRNC) associated with the LMU. Instructions concerning the timing, the nature, and any periodicity of these measurements are provided by the CRNC.

The location measurement unit may make its measurements in response to requests (e.g., from the CRNC), or it may autonomously measure and report regularly or when there are significant changes in radio conditions. A UE positioning request may involve measurement by one or more LMUs. The LMU may be of several types and the CRNC will select the appropriate LMU depending on the UE positioning method used.

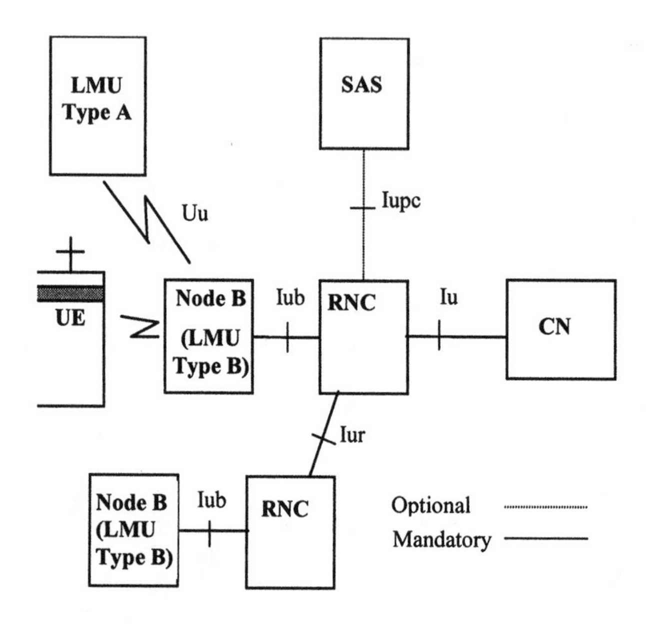

Figure 3.7 General architecture of terminal positioning in UTRAN.

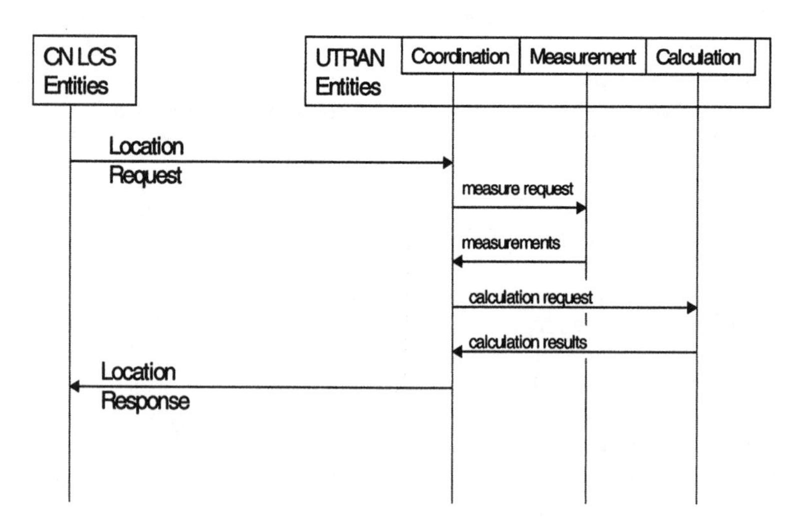

Figure 3.8 General sequence for UE positioning operation.

The LMU may be employed to measure UTRAN transmissions in the uplink or the downlink. These measurements may be made either to locate the UE or to measure a system parameter needed by the UE positioning. The measurements carried out by an LMU are generic and are usable by more than one positioning method. These include the following:

- Radio interface timing measurements. These cover UTRAN GPS timing of cell frames or SFN-SFN observed time difference of the signals transmitted by Node B, where timing differences are measured relative to either some common reference clock (UTRAN GPS timing of cell frames) or the signals of another Node B (SFN-SFN observed time difference).
- Intersystem timing measurements. These include timing measurements between the UTRAN radio signals transmitted by a Node B and an external system such as the GPS or GSM.

There are two classes of LMUs: the stand-alone LMU and the associate LMU. The stand-alone LMU communicates with RNCs via the Uu interface, whereas the associated LMU communicates with RNCs via the Iub interface. The associated LMU signaling protocol is the NBAP. The protocol for stand-alone LMU UTRAN signaling is the RRC protocol.

3.3.2.1 Stand-Alone LMU

A stand-alone LMU is similar to an ordinary UE and it is accessed exclusively over the UTRAN air interface, that is, the one between the mobile terminals and the base stations (Uu). There is no other connection from the stand-alone LMU to the UTRAN network.

A stand-alone LMU has a serving Node B that provides signaling access to its CRNC. A stand-alone LMU also has a serving 3G-MSC (mobile switching center), visitor location register (VLR), and a subscription profile in a home location register (HLR). A stand-alone LMU supports all radio resource and mobility management functions of the UTRAN radio interface that are necessary for signaling. A stand-alone LMU supports those connection management functions necessary to support UE positioning signaling transactions with the CRNC and may support certain call control functions to support signaling to a CRNC using a circuit-switched data connection.

To ensure that a stand-alone LMU and its associated CRNC can always access each other, an LMU may be camped on a particular cell site. For any stand-alone LMU with a subscription profile in an HLR, a special profile may be used to indicate the assigned supplementary services. An identifier in the HLR profile distinguishes an LMU from a normal UE. All other data specific to an LMU is administered in the LMU and in its associated CRNC.

3.3.2.2 Associated LMU

An associated LMU is accessed over the Iub interface from an RNC. An associated LMU may make use of the radio apparatus and antennas of its associated Node B. The LMU may be either a logically separate network element addressed using some pseudo cell ID or connected to or even integrated into a Node B. Signaling to an associated LMU is by means of messages routed through the controlling Node B. An associated LMU can be separated from the Node B, but still communicates with the CRNC via the Node B Iub interface.

3.3.3 Functions of Terminals

For LCS, the UE may transmit the needed signals in the uplink-based UE positioning measurements and/or make measurements of downlink signals. The measurements are determined by the chosen positioning method. The UE may also contain an independent positioning function, such as a GPS receiver, thus being able to report its position, independent of the UTRAN transmissions. The UE with an independent positioning function may also make use of the information broadcast by the UTRAN that assists the function.

The UE may also contain LCS applications or access an LCS application through communications with a network accessed by the UE. This application may include the needed measurement and calculation functions to determine the UE's position with or without assistance of the UTRAN UE positioning entities.

3.3.4 Stand-Alone SMLC

The serving mobile location center (SMLC) contains functionalities required to support LCS. The SMLC manages the overall coordination and scheduling of resources required to perform positioning of a UE. It also calculates the final location and estimates the accuracy. Physically, it can be a stand-alone unit, or be integrated into the serving RNC. In one public land mobile network (PLMN), there may be more than one SMLC.

A stand-alone SMLC (SAS) performs the following functions. First, for both UE-assisted and UE-based A-GPS positioning methods, it provides the GPS assistance data to the RNC, which is delivered to the UE through point-to-point or broadcast channels. Second, it acts as a location calculation server if the location estimates are not performed in the RNC.

3.4 CONCLUDING REMARKS

Mobile terminal positioning is one of the great challenges facing designers of future radio access networks. The cheapest, simplest, and most reliable method is based on cell ID. However, its accuracy is the lowest and it cannot meet the requirement of services such as emergency, tracking, and direction finding, although it is probably adequate for such applications as localized advertising. The GPS-based methods are the most accurate for outdoor services, but they may fail when the mobile terminal is inside a building or a canyon. Besides, they do have strong cost implication to terminals, as the terminals need to have built-in GPS receivers. It is foreseeable that when the indoor wireless local areas (WLAN) and wireless personal area networks (WPAN) become widespread and micro or pico cells are widely used to cover the traditional blind spots, the combination of Cell ID and A-GPS can offer an adequate positioning technique for most location-based services. The methods based on observed time difference of arrival (OTDOA) offer a good compromise among accuracy, reliability, and cost, but both the downlink OTDOA and uplink TDOA require a means of mitigating the multiuser interference problem in CDMA and the near-far problem in particular.

It should be pointed out that the position information of mobile terminals is very important not only for location services but also for the internal operation of radio access networks. In addition to assisting the mobile terminal handover between adjacent cells, for instance, the position information can be used to speed up the beamforming process in smart antennas and to optimize the radio resource management.

References

[1] Y. Zhao, "Standardization of Mobile Phone Positioning for 3G Systems," *IEEE Communications Magazine*, No. 7, Vol. 40, July 2002, pp. 108-116.

[2] J. C. Liberti and T. S. Pappaport, *Smart Antennas for Wireless Communications*, Upper Saddle River, NJ: Prentice Hall, 1999.

[3] J. J. Caffery and G. L. Stuber, "Subscriber Location in CDMA Cellular Networks," *IEEE Vehicle Technology*, Vol. 47, May 1998, pp. 406-416.

[4] J. J. Caffery and G. L. Stuber, "Overview of Radiolocation in CDMA Cellular Systems," *IEEE Communications Magazine*, Vol. 36, April, 1998, pp. 38-45.

[5] Cambridge Positioning Systems (CPS), Software Blanking for OTDOA Positioning, 3GPP *TSG RP*-020372, June 2002.

[6] TSGR1#4(99)346, Recapitulation of the IPDL Positioning Method, Ericsson, April 1999.

[7] TSGR1#8(99)g88, Evaluation of IP-DL Positioning Techniques Using Common Simulation Parameters, Ericsson, October 1999.

[8] http://www.cursor-system.com/cps/default.asp.

[9] http://www.trueposition.com.

[10] 3GPP TS 25.305, Functional Specification of UE Positioning in UTRAN.

[11] 3GPP TSG 23.271, V6.5.0, Functional Stage 2 Description of LCS.

[12] 3GPP TS 43.059, V6.1.0, Functional Stage 2 Description of Location Services (LCS) in GERAN.

APPENDIX 3A: OTDOA USING CIRCULAR VARIANT

The downlink OTDOA method requires a minimum of three spatially distinct base stations. All these base stations must be detectable by the mobile terminal whose position is to be determined. More than three measurements generally produce better location accuracy. The OTDOA positioning method relies upon measuring the time at which signals from the base station arrive at two geographically different locations – the mobile terminal itself and a fixed measuring point known as the location measurement unit (LMU), whose location is known. An implementation of the OTDOA method may require an LMU-to-node B ratio between 1:3 and 1:5. The position of the mobile terminal is determined by comparing the time differences between the two sets of timing measurements.

The mobile terminal performs measurements without the need for any additional hardware. For the circular variant, the arrival time from each of the base stations is measured individually at the UE and LMU. Based on the measured values, the location of the UE can be calculated either in the network or in the UE itself.

There are five parameters associated with the circular variant, which are defined in the following:

M_{ot} - The observed time at the UE at which a signal arrives from a base station. This is the time measured against the UE's internal clock.

L_{ot} - The observed time at the LMU at which a signal arrives from a base station. This is the time measured against the LMU's internal clock.

ε - The time offset between the UE's internal clock and the LMU's internal clock, which is an unknown variable.

D_{mb} - The geometrical distance from a UE to a base station.

D_{lb} - The geometrical distance from LMU to a base station.

These quantities are related by

$$D_{mb} - D_{lb} = v(M_{ot} - L_{ot} + \varepsilon) \tag{3A.1}$$

where v is the speed of light. There is one such equation for every base station. Since there are three unknown quantities (UE position x, y and clock offset ε), measurements must be made at least for three base stations in order to solve for the UE location and the unknown clock offset. The estimate of the position of the UE is defined by the intersection of circles centered on the base stations common to observations made by the UE and LMUs. The uncertainty associated with the radius of a circle is known as the measurement error margin. The overlap of the resulting areas is defined by a confidence ellipse, described by the length of its axes and its orientation.

Assuming the position of the mobile is defined by (x, y) and the measurements are performed against three base stations located at (x_i, y_i) with i = 1, 2, 3, one has

$$\sqrt{(x-x_1)^2+(y-y_1)^2}=v(T_{m1}-T_{l1}+\varepsilon)+D_{l1} \tag{3A.2a}$$

$$\sqrt{(x-x_2)^2+(y-y_2)^2}=v(T_{m2}-T_{l2}+\varepsilon)+D_{l2} \tag{3A.2b}$$

$$\sqrt{(x-x_3)^2+(y-y_3)^2}=v(T_{m3}-T_{l3}+\varepsilon)+D_{l3} \tag{3A.2c}$$

Let

$$z=v\varepsilon \tag{3A.3a}$$

$$E_1=v(T_{m1}-T_{l1})+D_{l1} \tag{3A.3b}$$

$$E_2=v(T_{m2}-T_{l2})+D_{l2} \tag{3A.3c}$$

$$E_3=v(T_{m3}-T_{l3})+D_{l3} \tag{3A.3d}$$

One has

$$\sqrt{(x-x_1)^2+(y-y_1)^2}=z+E_1 \tag{3A.4a}$$

$$\sqrt{(x-x_2)^2+(y-y_2)^2}=z+E_2 \tag{3A.4b}$$

$$\sqrt{(x-x_3)^2+(y-y_3)^2}=z+E_3 \tag{3A.4c}$$

which can be expanded as follows:

$$x^2-2x_1x+x_1^2+y^2-2y_1y+y_1^2=z^2+2E_1z+E_1^2 \tag{3A.5a}$$

$$x^2-2x_2x+x_2^2+y^2-2y_2y+y_2^2=z^2+2E_2z+E_2^2 \tag{3A.5b}$$

$$x^2-2x_3x+x_3^2+y^2-2y_3y+y_3^2=z^2+2E_3z+E_3^2 \tag{3A.5c}$$

Let

$$A=2\begin{bmatrix} x_2-x_1 & y_2-y_1 & E_2-E_1 \\ x_3-x_2 & y_3-y_2 & E_3-E_2 \\ x_3-x_1 & y_3-y_1 & E_3-E_1 \end{bmatrix} \tag{3A.6}$$

$$V = \begin{bmatrix} (x_2^2 - x_1^2) + (y_2^2 - y_1^2) + (E_1^2 - E_2^2) \\ (x_3^2 - x_2^2) + (y_3^2 - y_2^2) + (E_2^2 - E_3^2) \\ (x_3^2 - x_1^2) + (y_3^2 - y_1^2) + (E_3^2 - E_1^2) \end{bmatrix} \tag{3A.7}$$

One obtains the following solution:

$$\overline{P} = A^{-1}V \tag{3A.8}$$

where

$$\overline{P} = [x \quad y \quad v\varepsilon]^T \tag{3A.9}$$

with "T" denoting the transpose. x and y in (3A.9) define the position of the mobile terminal.

Chapter 4

High-Speed Downlink Packet Access

Compared with the second generation mobile communications systems, UTRAN Release 99 provides enhanced packet data services with the data rates of 384 kbps for wide area coverage and up to 2 Mbps for hot spots. To meet the ever-increasing demand on data-centric applications such as multimedia and streaming services, a new technology enhancement in UTRAN, high-speed downlink packet access (HSDPA), has been introduced to the Release 5 specification of 3GPP as the next major evolution step [1–3].

The HSDPA technology will increase the UTRAN network capacity, reduce the round-trip delay, and increase the peak data rates up to 14 Mbps. To achieve these goals, a new shared downlink channel, known as the high-speed downlink shared channel (HS-DSCH), is employed to serve as a downlink "fat pipe." Its corresponding physical channel, the high-speed physical downlink shared channel (HS-PDSCH), employs a set of spreading codes with a fixed spreading factor of 16. Depending on the user equipment (UE) capability, an HSDPA user can use up to 15 codes in the set, which is equivalent to having a spreading factor close to 1 [4]. By dynamically assigning the HS-PDSCH resource to different users in a time-multiplexed manner, unnecessary idle periods can be avoided and the maximum trunking gain can be achieved. In addition, three fundamental techniques, which are tightly coupled with each other and rely on the rapid adaptation of the transmission parameters to the instantaneous radio conditions, are employed to increase the throughput of the HS-PDSCH channel: adaptive modulation and coding (AMC), hybrid automatic-repeat-request (HARQ), and fast scheduling.

The adaptive modulation and coding technique enables the use of spectral-efficient higher-order modulation schemes such as 16QAM [5] when channel conditions permit, and reverts to the robust QPSK modulation in less favorable channel conditions. Also, the number of redundancy bits and therefore the coding rates are adapted to the variation of the channel quality. The HARQ technique dynamically integrates the two powerful error control schemes: forward error correction (FEC) and automatic-repeat-request (ARQ). The receiver rapidly requests the retransmission of more data if a delivered data block is found erroneous, and the receiver combines the information from the original transmission and any subsequent retransmissions to decode a message. Last but not least, fast scheduling of users sharing the HS-DSCH ensures that the time-varying radio channel conditions for mobile terminals in different places of the cell are instantaneously exploited to achieve maximum throughput in the downlink while maintaining a certain fairness.

In this chapter, the fundamental principles of HSDPA and its operation in UTRAN are explained. The radio resource and mobility management issues are discussed. The impact of HSDPA on both terminals and the UTRAN architecture is studied and some implementation and deployment issues are addressed.

4.1 FUNDAMENTAL PRINCIPLES

4.1.1 Adaptive Modulation and Coding

One of the major techniques introduced to the early CDMA cellular systems is power control. The idea is to increase the transmission power when the quality of the received signal is poor and decrease the transmission power when the quality of the received signal exceeds a given threshold. This results in reliable communications between the transmitter and the receiver. Also, since the power control technique reduces the unnecessary intercell and intracell interference caused by excessive transmit power, the overall system capacity is increased.

An alternative technique to power control in dealing with the time-varying effect of the wireless channel is to "ride" the fading profile of the channel. Instead of trying to keep the signal quality at the receiver constant, one can change the modulation and coding scheme of the transmitted signal in such a way that more information-bearing bits are sent when the channel condition is good, and less information-bearing bits are transmitted when the channel condition deteriorates. This technique is known as adaptive modulation and coding (AMC), or link adaptation. AMC is especially suited for packet communications with a bursty nature, as the transmission can be scheduled only at the constructive fades of the channel [5]. In a cellular environment, the base station antennas are fixed, but the channel conditions for different mobile terminals change with time and location. Statistically, there should always be some mobile terminals enjoying good channel conditions and some suffering from poor channel conditions. Therefore, it should be possible for the network to send data to some of the mobile terminals in a cell with high speed by using higher modulation and higher rate coding schemes without increasing the transmit power. Compared with the conventional power control technique, AMC can lead to much higher system capacity for packet radio systems.

The adaptive modulation and coding technique is aimed at changing the modulation and coding format in accordance with variations in the channel conditions. The channel conditions can be estimated based either on the feedback from the receiver or the transmission power of other downlink channels under power control. In a system with AMC, users in favorable positions, such as the ones close to the cell site or at the peak of a fading profile, are typically assigned a higher-order modulation with higher code rates, such as 64QAM/16QAM and 3/4 rate Turbo codes, whereas users in unfavorable positions, such as the ones close to the cell boundary or at the lower peaks of a fading profile, are assigned a lower-order modulation with lower coding rates, such as QPSK with 1/2 rate Turbo codes (see Figure 4.1). If there are a large number of users in cell and the channel conditions of different users vary with time, which is normally the case, the base station can choose to serve users in favorable conditions and use high modulation schemes and coding rates most of the time, and the system capacity would be greatly improved. Another advantage of the AMC is that, since the transmit power is fixed (no fast power control is used), the interference to other users is significantly reduced.

In summary, the main benefits of the adaptive modulation and coding technique are the following:

- Higher data rates can be achieved for users in favorable conditions, which in turn increases the average throughput of the cell.
- Interference is reduced due to link adaptation based on variations in the modulation and coding scheme instead of variations in the transmit power.

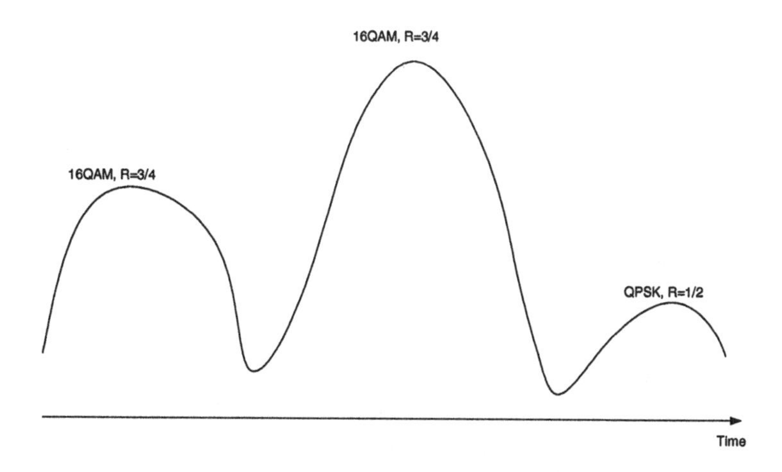

Figure 4.1 Illustration of the channel fading and adaptive modulation and coding, where R represents the coding rate.

For HSDPA, 3GPP Release 5 specifies 2-ms frame length. This ensures that, especially for moving HSDPA mobile terminals, there is enough time for the network to take advantage of the constructive fades and to perform retransmissions if necessary. It should be mentioned that, by average, higher modulation schemes inevitably require higher transmit power from the transmitter than lower modulation schemes in order to achieve the same bit error rate and to reach the same distance. In a cellular environment, however, the transmit power for each cell is limited and it is shared by different downlink channels. Given such a constraint, the benefit of using 64QAM as opposed to using 16QAM may be limited. Besides, 64QAM requires more complicated and thus more costly terminals. Therefore, the modulation schemes specified for HSDPA are QPSK and 16QAM. For illustration, Figure 4.2 shows the constellation of 16QAM specified in [6].

4.1.2 Hybrid ARQ

A major task in data communications is to control transmission errors caused by the channel noise and interference so that the data can be delivered to the user correctly. There are two basic error-control strategies: forward error correction (FEC) and automatic-repeat-request (ARQ) [7].

In forward error correction (FEC), the information bearing bits are protected by some coding scheme employing redundancy bits, and most errors caused by channel impairment can be recovered at the receiver by the decoder. One advantage of the forward error correction is that no feedback channel is required and therefore the delay is only contained in the decoding process. The disadvantage of the forward error correction scheme, however, is that the decoded data is always delivered to the user regardless of whether it is correct or incorrect. In order to achieve high reliability, a long and powerful error-correcting code must be used and a large number of error patterns must be corrected. This means that a stand-alone forward error correction scheme normally has a low coding rate, which is defined as the ratio between the number of information-bearing bits and the total number of bits transmitted. Unfortunately, a low coding rate results in inefficient use of the spectrum in good channel conditions.

In automatic-repeat-request (ARQ), high-rate error detection codes are normally used and a retransmission is requested if the received data is found to be erroneous [7, 8]. Using a proper error-correction code, the probability of undetected errors can be made reasonably small and the final error-free data delivery is left as the task of ARQ. ARQ schemes are widely used in data communications systems, as they are simple and provide high system reliability. The drawbacks of the ARQ schemes, however, are that the throughput may be not constant and it falls rapidly with increasing bit error rates.

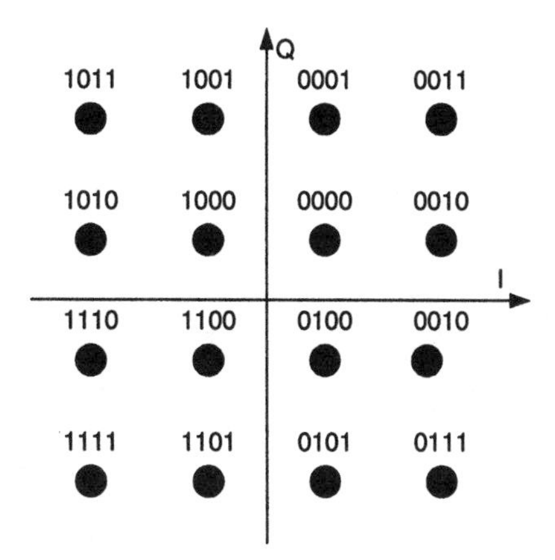

Figure 4.2 An illustration of a 16QAM constellation.

4.1.2.1 ARQ

In a practical system, the functionalities of the ARQ include the following [7]:

- Error detection. The receiver detects errors in the so-called protocol data unit (PDU).
- Positive acknowledgment. The receiver returns a positive acknowledgment (ACK) to the successfully received, error-free PDUs.
- Retransmission after timeout. The transmitter retransmits the PDU that has not been acknowledged after a predetermined amount of time.
- Negative acknowledgment and retransmission. The receiver returns a negative acknowledgment (NAK) to the erroneous PDUs and the transmitter retransmits them.

Admittedly, not all the procedures described above are used in every ARQ scheme.

Broadly speaking, there are three types of basic ARQ schemes: stop-and-wait (SAW), go-back-N, and selective-repeat (SR) [7]. In a stop-and-wait ARQ system (see Figure 4.3), the transmitter sends off a packet and then waits for an acknowledgment. A positive acknowledgment (ACK) indicates that the packet has been delivered successfully, and the transmitter then sends the next packet in the queue. A negative acknowledgment (NAK) from the receiver indicates that at least one error has been detected in the packet and the transmitter then resends the packet and waits for further acknowledgment. Stop-and-wait is the simplest scheme but inherently inefficient because of the idle time spent waiting for the acknowledgment of each transmitted packet before the transmitter starting the next transmission. In a slotted system in which the transmission of data packets is aligned with some predetermined time slots, the feedback delay will waste at least half of the system capacity, as the transmitter needs to wait for the acknowledgments from the receiver. As a result, at least every other time slot must go idle even on an error-free channel. However, the stop-and-wait ARQ does have some advantages. First, it requires very little overhead. In a stop-and-wait scheme, correctness is ensured with a simple 1-bit sequence number that identifies the current or the next block. As a result, the control overhead is minimal. The acknowledgment overhead is also minimal, as the indication of a successful/unsuccessful decoding (using ACK and NAK) may be signaled concisely with a single bit. Furthermore, because only a single block is in transit at a time, memory requirement at the receiver is also minimized.

Denoting p as the probability of packet loss, R as the data transmission speed, T_i as the idle time of the transmitter between two successive transmissions, k as the number of information bits in the data block, and n as the total number of bits in the whole data block, the throughput of a stop-and-wait (SAW) system is given by

$$Throughput_{SAW} = k\frac{1-p}{n+RT_i} \tag{4.1}$$

In (4.1), RT_i represents the wasted capacity due to idle periods and it can be seen that the greater the data transmission rate and the idle period, the lower the throughput. Further, if the data transmission rate is increased, the idle period must be decreased to keep the throughput constant, but this could be difficult, as T_i is determined by the propagation and processing delay.

In contrast to the stop-and-wait protocol, selective-repeat (SR) is the most theoretically efficient ARQ scheme among the three basic ones and it has been employed by many systems, including the Release 99 UTRAN radio link control (RLC) layer. As shown in Figure 4.4, its basic principle is only to retransmit the negatively acknowledged PDUs. For an ideal SR system in which the receiver has an unlimited buffer size, the throughput is given by

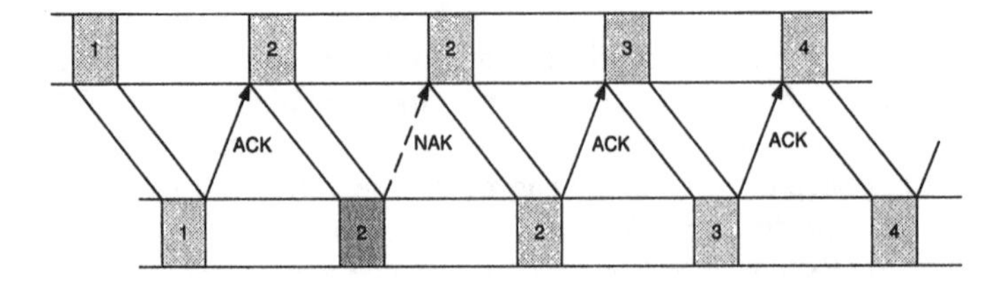

Figure 4.3 Illustration of the stop-and-wait ARQ.

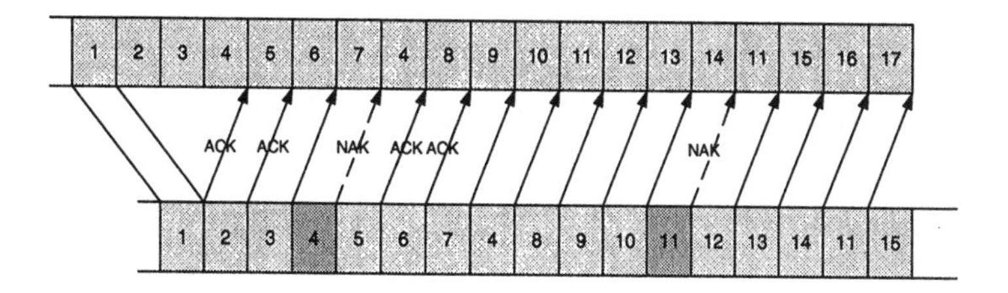

Figure 4.4 Illustration of the selective repeat ARQ.

$$Throughput_{SR} = k\frac{1-p}{n} \tag{4.2}$$

Compared with (4.1), it is seen that the throughput of the selective-repeat ARQ is independent of the transmission rate and the feedback delay, as no waiting time is needed.

SR is theoretically insensitive to delay and has the favorable property of repeating only those blocks that have been received in error. To accomplish this task, however, the SR ARQ transmitter must employ a sequence number to identify each block that it sends. SR may fully utilize the available channel capacity by ensuring that the maximum block sequence number (MBSN) exceeds the number of blocks transmitted in one round-trip feedback delay. The greater the feedback delay, the larger the maximum sequence number must be.

Unfortunately, there is a major problem in the implementation of selective-repeat ARQ in HSDPA – high memory requirement in the mobile terminal. This is due to the fact that the mobile terminal must store soft samples for each transmission of a block. Since MBSN blocks may be in transit at any time, a large MBSN requires significant storage in the mobile terminal, which consequently increases the terminal cost. For this reason, only the multichannel stop-and-wait scheme is employed in HSDPA.

4.1.2.2 Combining ARQ with FEC

Hybrid ARQ schemes combine the conventional automatic retransmission request (ARQ) with forward error coding (FEC) and they can be regarded as an implicit link adaptation technique [9, 10]. A hybrid ARQ system consists of an FEC subsystem contained in an ARQ system. The function of the FEC system is to reduce the frequency of retransmission by correcting the error patterns that occur most frequently, thus ensuring a high system throughput. When a less frequent error pattern occurs and is detected, the receiver requests a retransmission instead of passing the erroneous data to the user. This increases the system reliability. Hybrid ARQ (HARQ) schemes offer the potential of better performance and lower implementation cost if appropriate ARQ and FEC schemes are properly combined. In the following, the operation principles of the three types of HARQ schemes, HARQ Type I, HARQ Type II, and HARQ Type III, are presented briefly [4, 5], and the pros and cons of each scheme are analyzed.

In HARQ Type I, the same protocol data unit (PDU) is retransmitted until the receiver accepts it as error-free or until the maximum number of allowed retransmission attempts is reached. HARQ Type I scheme is best suited for communications systems in which a fairly constant level of noise and interference is anticipated in the channel. In this case, an adequate amount of redundancy can be built into the system to correct the vast majority of errors so the number of retransmissions can be kept to a minimum. For nonstationary channels, such as the ones encountered in mobile communications, the HARQ Type I can be very inefficient. When the channel condition is good, the large number of redundancy bits built into the FEC subsystem result in a waste of bandwidth. On the other hand, when the channel condition is poor, the FEC may be not powerful enough and too many retransmissions may be needed.

HARQ Type II, or adaptive hybrid ARQ, is particularly suited for time-varying channels. In this scheme, the concept of incremental redundancy is employed and the received PDUs are concatenated to form corrupted code words from increasingly longer and lower rate codes. In the first transmission, the PDU may be coded with a high-rate code (low redundancy) for error detection and correction. If the receiver detects the presence of errors in the PDU, it saves the erroneous PDU in a buffer and at the same time requests a retransmission. Unlike HARQ Type I, what is retransmitted in HARQ Type II is not the original PDU but a block of new data. The new data are formed based on the original PDU and the error correcting code used. When the new PDU is received, it is used to correct the errors in the erroneous PDU stored previously in the buffer. If the second attempt fails again, the receiver will request a further retransmission and this process continues until satisfactory results are achieved.

HARQ Type III also belongs to the class of incremental redundancy ARQ schemes. With HARQ Type III, however, each retransmission is self-decodable. Chase combining (also called HARQ Type III with one redundancy version) involves the retransmission by the transmitter of the same coded data packet. The decoder at the receiver combines these multiple copies of the transmitted packet weighted by the signal-to-noise ratio (SNR) of the received signal for each attempt. As a result, diversity gain in the time domain is obtained and the temporal diversity may also include contributions of spatial diversity implicitly. Compared with HARQ Type II, HARQ Type III is much easier to implement.

In the UTRAN HSDPA specification, three HARQ schemes are specified, namely, Chase combining, full incremental redundancy, and partial incremental redundancy. In full incremental redundancy, the initial transmission carries the information bits and some redundancy bits. If the decoding fails, more redundancy bits are sent in each subsequent retransmission until the data frame is correctly decoded. It can be easily seen that the memory needed at the terminal for full incremental redundancy depends on the number of retransmissions, and therefore is variable. In partial incremental redundancy, the original information bits and some redundancy bits are sent in each transmission. The content of the redundancy bits vary in different transmissions, but the data block in each transmission is self-decodable. If decoding fails in any particular attempt, the received data block can be combined with the preceding ones to increase the coding power. The performance of different type of HARQs depends on the condition of the radio channel. Considering all factors, including spectral efficiency, implementation complexity, and robustness, it appears that Chase combining offers a very attractive compromise. In UTRAN, it is the scheduler that determines the redundancy version parameters for the HARQ functional entities in node Bs.

The choice of ARQ mechanism is important. As discussed earlier, the selective-repeat ARQ is most efficient, but it requires high memory in the mobile terminal. Furthermore, HARQ requires the receiver to reliably determine the sequence number of each transmission. Unlike conventional ARQ, every block is used in HARQ even if there is an error in the data. In addition, the sequence information must be very reliable to sustain poor channel conditions that are the cause of errors in the data. Typically, a separate, strong code is needed to encode the sequence information and this increases the required signaling bandwidth.

The stop-and-wait does have the salient advantage in the ease of implementation. Its greatest disadvantage, low efficiency, can be overcome by employing some parallel processes to keep the channel busy most of the time. In fact, the so-called *N* channel stop-and-wait HARQ has been adopted by 3GPP for UTRAN. In essence, it runs a separate instantiation of the Hybrid ARQ protocol when the channel is idle in order to increase throughput. When one instance of the algorithm deals with a data block on the downlink, another instance of the algorithm sends an acknowledgment on the uplink. The roles of the instances rotate in a round-robin manner. The number of instantiations increases with the delay between the transmission of the data block and the reception of the acknowledgment. The drawback of this scheme is that the receiver needs to store *N* blocks of data. In a typical cellular environment, the delay from the transmission of the data block to the reception of acknowledgment is no more than eight HSDPA TTIs (16-ms). Therefore, 3GPP specifies that the maximum number of the stop-and-wait HARQ processes is eight. Each HARQ process is a self-contained channel on which a stop-and-wait (SAW) HARQ protocol is operated.

The adaptive modulation and coding (AMC) scheme provides some flexibility in choosing an appropriate modulation and coding scheme for the channel conditions based on measurements either in the UE or in the network. The disadvantage of the AMC is that it demands an accurate and timely measurement, but certain system delays caused by the feedback loop are inevitable. In contrast, the HARQ autonomously adapts to the instantaneous channel conditions and is insensitive to the measurement error and delay. Combining the two complementary techniques, AMC and HARQ, leads to an integrated robust and high-performance solution, in which AMC provides the coarse data rate selection, whereas the HARQ provides for fine data rate adjustment based on channel conditions.

4.1.3 Fast Scheduler

As a central coordination unit for the transmission and retransmission of data blocks targeted at different mobile terminals in a cell, the scheduler is a key element in the operation of HSDPA. To large extent, it determines the overall behavior of the system. The scheduler exploits the multiuser diversity and strives to transmit to users whose radio conditions permit high data rates. In the meantime, it must also maintain a certain degree of fairness. Generally speaking, a greater allowable variance in the quality of services (QoS) results in greater system capacity, whereas a tighter QoS requirement normally leads to a reduced system capacity. The scheduler must be properly designed to make use of different channel conditions and to meet the QoS requirements of different service classes. Since 3GPP has decided not to standardize the scheduling algorithm, different schedulers can be designed by network vendors to suit the requirements of operators and environments. Information on which the scheduler can base its decisions includes the estimated channel quality as used by the link adaptation, the current load of the cell, and traffic priority classes.

The scheduler for HSDPA is referred to as being fast due to the fact that, compared with Release 99 specifications, the scheduler is moved from RNC to node Bs to reduce delays so faster scheduling decisions can be made. In addition to other functionalities, such as the choice of redundancy version and modulation and coding scheme, a fundamental task of the scheduler for HSDPA is to schedule the transmission for users. As shown in Figure 4.5, data to be transmitted to users are placed in different queues in a buffer and the scheduler needs to determine the sequential order in which the data streams are sent. The simplest scheduling algorithm is the round-robin method, which selects the user packets in a round-robin fashion. In this method, the number of time slots allocated to each user can be chosen to be inversely proportional to the users' data rates, so the same number of bits are transmitted for every user in a cycle. Obviously, this method is the "fairest" in the sense that the average delay and throughput would be the same for all users. Alternatively, one can fix the number of time slots allocated to each user, so the same time is spent on the transmission to every user. Admittedly, this does not lead to the same user throughput. However, there are two disadvantages associated with the round-robin method. The first is that it disregards the conditions of the radio channel for each user, so users in poor radio conditions may experience low data rates, whereas users in good channel conditions may not even receive any data until the channel conditions turn poor again. This is obviously against the spirit of the HSDPA and it would lead to the lowest system throughput. The second disadvantage of the round-robin scheduler is that there is no differentiation in the quality of services for different classes of users [11].

Another extreme scheme is called the maximum carrier-to-interference (C/I) ratio method, or Max C/I. In this method, the scheduler attempts to take advantage of the variations in the radio channel conditions for different users to the maximum, and always chooses to serve the user experiencing the best channel condition, that is, the one with maximum carrier-to-interference ratio. Apparently, the max C/I scheduler leads to the maximum system throughput but is the most unfair, as users in poor radio conditions may never get served or suffer from unacceptable delays.

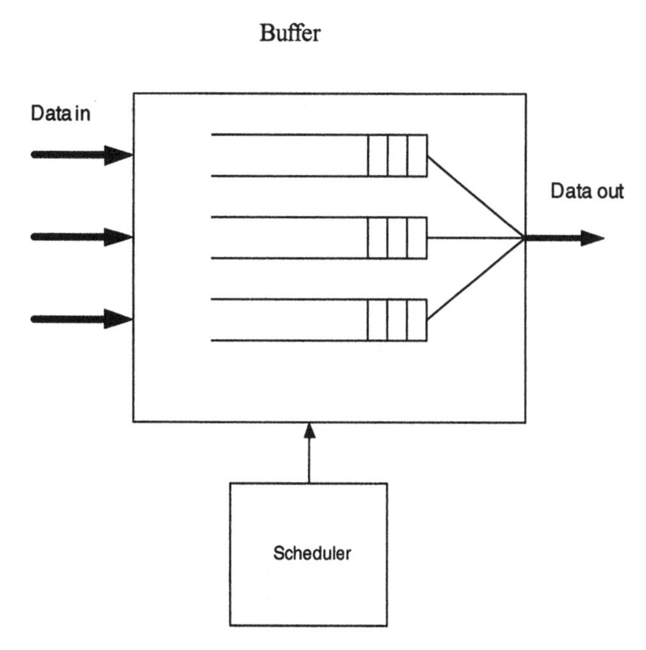

Figure 4.5 Illustration of scheduling.

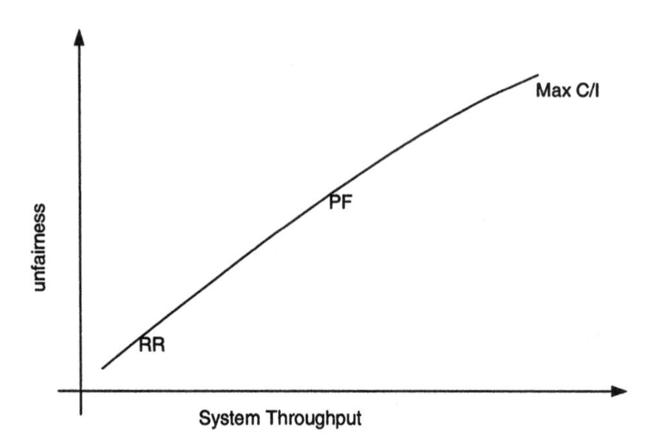

Figure 4.6 Performance of different scheduling algorithms.

A practical and balanced scheduling method is the proportional fairness scheduling adopted by CDMA2000 1xEV-DO [12]. It takes into account both the short-term variation of the radio channel conditions and the long-term throughput of each user. In this method, the user with the largest $R[n]/R_{av}$ is served first, where $R[n]$ is the data rate in the current time slot n and R_{av} is the average data rate for the user in the past average window. The size of the average window determines the maximum duration that a user can be starved from data, and as such it reflects the compromise between the maximum tolerable delay and the cell throughput. According to this scheduling scheme, if a user is enjoying a very high average throughput, its $R[n]/R_{av}$ will probably not be the highest. Then it may give way to other users with poor average throughput and therefore high $R[n]/R_{av}$ in the next time slots, so the average throughput of the latter can be improved. On the other hand, if the average throughput of a user is low, the $R[n]/R_{av}$ could be high and it might be granted the right of transmission even if its current channel condition is not the best.

Figure 4.6 illustrates the performance of different scheduling algorithms. It is shown that the proportional fairness scheduling algorithm offers a good compromise between fairness viewed by each user and the overall system throughput. Unfortunately, the proportional fairness algorithm as it stands cannot fully meet the QoS requirements of different users fully. A further improvement to the algorithm could be to add extra constraints. For instance, the criterion for choosing the next data block to transmit can be changed from maximum $R[n]/R_{av}$ to maximum $R[n]/R_{av}$ subject to certain QoS constraints.

Fast scheduling and AMC, in conjunction with HARQ, is a way of maximizing the instantaneous use of the fading radio channel in order to realize maximum throughput. The HSDPA technology enables higher-rate data transmission through a higher-modulation and coding rate and limited retransmissions, while keeping the power allocated to HS-DSCH channel in a cell constant. Notwithstanding, the slow power control is still needed to adjust the power sharing among terminals and between different channel types.

4.2 HS-DSCH AND ASSOCIATED CHANNELS

In HSDPA, a new transport channel, HS-DSCH, which is primarily designed for best-effort packet data, is introduced. Similar to the downlink shared channel (DSCH) in UTRAN Release 99 specifications, the HS-DSCH employs a set of channelization (spreading) codes shared by the active HSDPA mobile terminals in a cell. The HS-DSCH is allocated to users mainly on the basis of the transmission time interval (TTI), in which users are allocated within different TTIs. To reduce delays, increase the granularity in the scheduling process, and facilitate better tracking of the time-varying channel conditions, HS-DSCH employs 2-ms TTI, which is much shorter than the minimum 10-ms TTI specified for Release 99 channels. The HS-DSCH code resource consists of up to 15 channelization codes with a spreading factor of 16. The primary way of sharing this resource is in the time domain, although it is also permissible to share the resource in the code domain. The physical channel carrying HS-DSCH transport traffic is termed HS-PDSCH, and each HS-PDSCH is identified by its specific channelization code. Therefore, there can be up to 15 HS-PDSCHs in a cell. Figure 4.7 illustrates the slot structure of an individual HS-PDSCH subframe and it is seen that there are three timeslots in each TTI. In Figure 4.7, M is the number of bits per modulation symbol, that is, $M = 2$ for QPSK and $M = 4$ for 16QAM. This means that for HS-PDSCH, both QPSK and 16QAM modulation schemes can be used and these modulation schemes lead to different slot formats. The corresponding data rates for QPSK and 16QAM are shown in Table 4.1, where SF refers to the spreading factor [13]. It is shown that, with 16QAM, a single HS-PDSCH channel can achieve a data rate of 960 kbps. Using 15 HS-PDSCH channels (codes), HSDPA can produce the headline data rate of 14.4 Mbps.

Table 4.1

HS-PDSCH Data Rates

Modulation	*Channel Bit Rate (Kbps)*	*Channel Symbol Rate (Kbps)*	*SF*
QPSK	480	240	16
16QAM	960	240	16

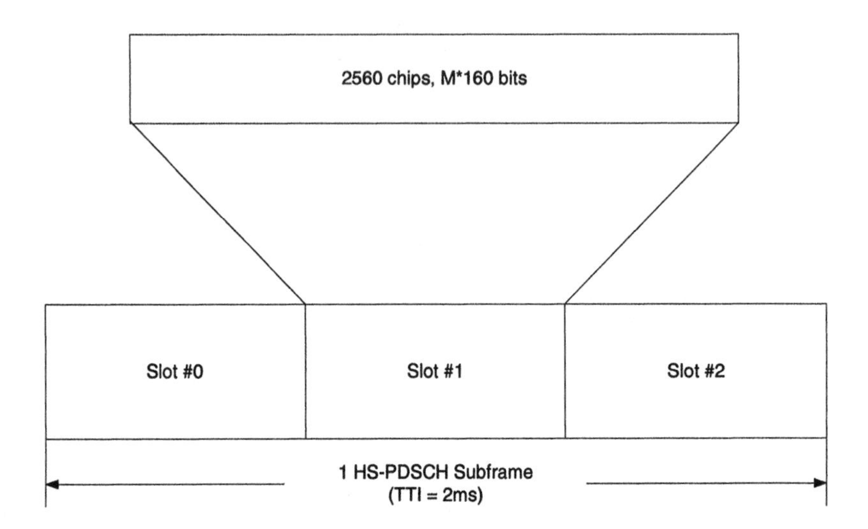

Figure 4.7 Subframe structure of the HS-PDSCH, where M = 2 for QPSK and M = 4 for 16QAM.

The HSDPA specification does permit simultaneous transmissions. For instance, two to four users can be supported within the same TTI by using different subset of the channelization codes allocated to HS-DSCH. Figure 4.8 gives an example of supporting three users, where two HSDPA TTIs are shown. It is seen that in the first time slot five HS-PDSCH channels are allocated to UE #1. In the second and third time slots, two HS-PDSCH channels are allocated to UE #2 and three HS-PDSCH channels are allocated to UE #3.

HS-SCCH

Besides user data, the node B must also transmit associated control signaling to user terminals, so terminals scheduled for the upcoming HS-DSCH TTI can be notified. Similarly, additional lower-layer control information such as the transport format, including the modulation and coding schemes to be used, and hybrid ARQ related information must be transmitted. This control information applies only to the user equipment that is receiving data on the HS-DSCH and is transmitted on a shared control channel, HS-SCCH. This means that the HS-PDSCH does not carry any physical layer information.

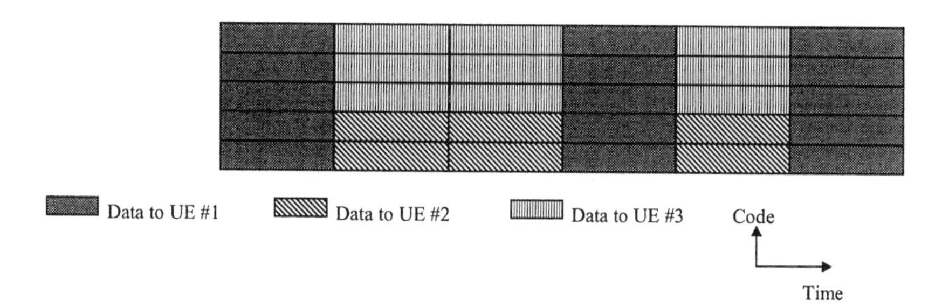

Figure 4.8 Illustration of the HS-PDSCH supporting three users.

DL-DPCH & UL-DPCH

As with the DSCH in Release 99, every mobile terminal to which data can be transmitted on the HS-DSCH has an associated downlink dedicated physical channel (DL-DPCH). The DL-DPCH is used to carry power control commands for the associated uplink (UL-DPCH) and, if needed, for other services such as circuit-switched voice. A special use of the DL-DPCH for HSDPA, however, is that an HSDPA indicator (HI) is transmitted. The HI consists of two information bits indicating which HSDPA shared control channel (HS-SCCH) subset the terminal should monitor in order to obtain the signaling information. The UL-DPCH carries power control signals for the DL-DPCH and this information, together with the feedback information provided on the HS-DPCCH, can be used by the network to estimate the channel quality more accurately.

HS-DPCCH

In addition to downlink signaling, uplink control signaling is also required for such purposes as the HARQ protocol status report, including ACK and NAK, and the measurement report on the quality of the radio channel in which the UE is operating. This information is carried on a separate code multiplexed uplink dedicated physical control channel, HS-DPCCH, which is invisible to Release 99 and Release 4 node Bs. There is one HS-DPCCH for each active terminal using HSDPA services.

Figure 4.9 shows the physical channels associated with HSDPA users and the roles they play in the HSDPA operation. In the uplink, there are the uplink dedicated physical channel (UL-DPCH) for power control and the HS-DPCCH for feedback information from the UE. In the downlink, there are HS-PDSCH for HSDPA data traffic, HS-SCCH for signaling and the DL-DPCH for HSDPA indicator (HI). Figure 4.10 shows the relative timing between the HS-SCCH and the associated HS-PDSCH for one HS-DSCH subframe. It is seen that the HS-PDSCH starts $\tau_{HS\text{-}PDSCH} = 2\times T_{slot} = 5{,}120$ chips after the start of the HS-SCCH. This is designed to give the terminal enough time to obtain the physical layer information before starting to decode the data on HS-PDSCH.

The HS-PDSCH power control is managed by node Bs. When the HS-PDSCH is transmitted using 16QAM, the transmission power is kept constant during the corresponding HS-DSCH subframe (or TTI). For the case of multiple HS-PDSCH transmission to one UE, all the HS-PDSCHs intended for that UE are allocated with the same power level. The sum of the powers used by all HS-PDSCHs and HS-SCCHs in a cell is controlled by the parameter, *HS-PDSCH and HS-SCCH Total Power,* which is signaled by the RNC.

- Upon the initial connection on dedicated channels, the following specific parameters are informed to the UE by RNC via higher layer signaling:
- The HS-SCCH set to be monitored.
- Repetition factor of ACK/NAK : *N_acknak_transmit.*
- Channel quality indicator (CQI) feedback cycle k, which has a possible value of [1, 5, 10, 20, 40, 80] that corresponds to the feedback cycle of [2, 10, 20, 40, 80, 160] ms. In addition, if $k = 0$, the measurement feedback can be shut off completely.
- CQI feedback offset l.
- Repetition factor of CQI: *N_cqi_transmit.*
- The HS-SCCH set to be monitored.
- Repetition factor of ACK/NAK : *N_acknak_transmit.*

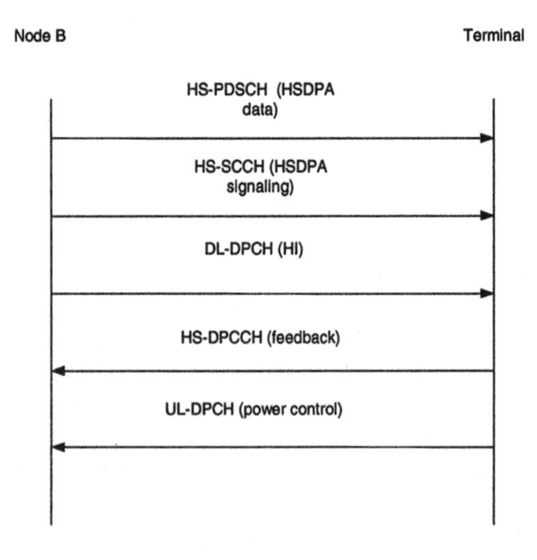

Figure 4.9 Channels involved in the HSDPA operation.

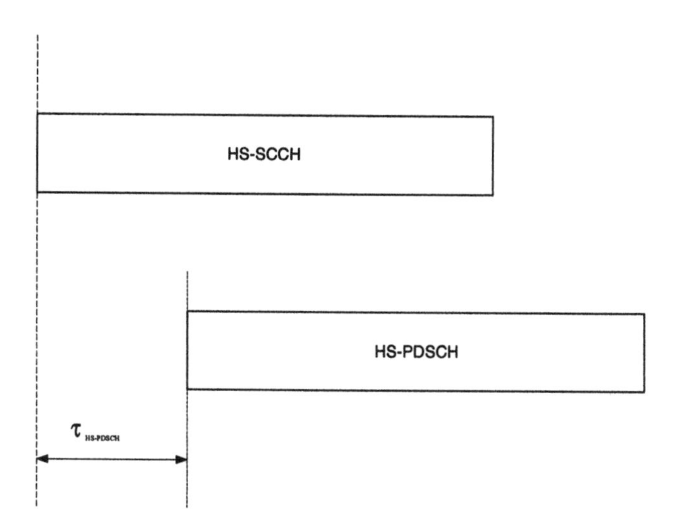

Figure 4.10 Timing relation between the HS-SCCH and the associated HS-PDSCH.

The use for the measurement feedback cycle k and feedback offset l is illustrated in Figure 4.11.

4.2.1 Coding for HS-DSCH Data Block

The HSDPA data arrives at the coding unit in the form of a maximum of one transport block in every transmission time interval (TTI). The transmission time interval is 2 ms, which is mapped to a radio subframe of three time slots. As shown in Figure 4.12, the following coding steps are to be performed on the HS-DSCH data block [13]:

- Add cyclic redundancy check (CRC) bits as attachment to each transport block.
- Bit scrambling. The bits output from the above operation are scrambled in the bit scrambler.
- Code block segmentation.
- Channel coding.
- Hybrid ARQ processing.
- Physical channel segmentation.
- Interleaving for HS-DSCH.
- Constellation rearrangement for 16QAM (if applicable).
- Mapping to physical channels.

Some of these operation steps are self-explanatory and others are explained in the following.

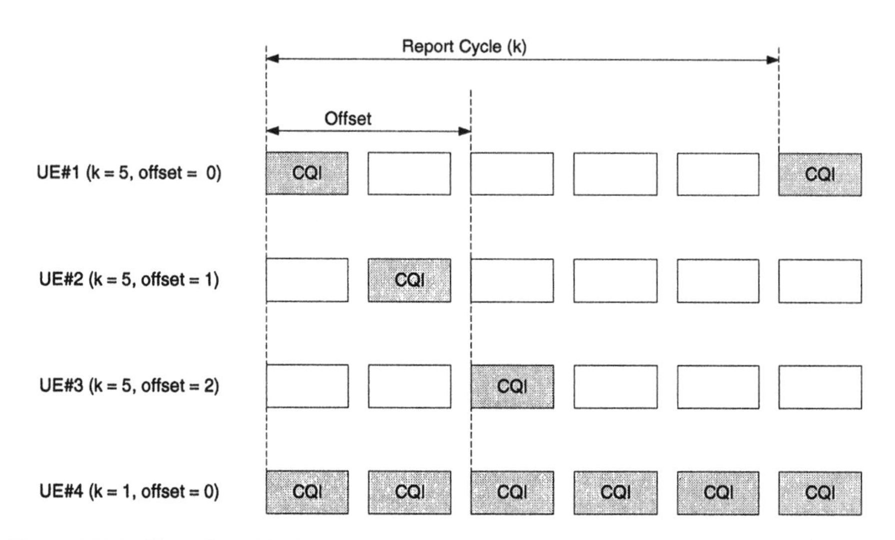

Figure 4.11 An illustration of feedback measurement transmission timing.

Cyclic redundancy check (CRC) is one of the most common and powerful error-detection coding schemes used in data communications. Its principle can be described as follows. Given a k-bit data block, the encoder generates an n-bit sequence, known as the frame check sequence, so that the resulting frame consisting of $n+k$ bits is exactly divisible by some predetermined sequence. The receiver uses the sequence then to divide the incoming data frame. If there is no remainder at the receiver, it assumes that the frame has been correctly received. Otherwise, the received frame is declared erroneous.

When more than one HS-PDSCH is used, the physical channel segmentation function distributes the bit stream among different physical channels. For convenience, denoting R as the number of bits input to the physical channel segmentation block, P as the number of physical channels employed, and U as the number of bits in one radio subframe for each HS-PDSCH, one has

$$U = \frac{R}{P} \tag{4.3}$$

The interleaving for HS-PDSCH is done separately for each physical channel, as shown in Figure 4.13. The bits input to the block interleaver are denoted by $u_{p,1}, u_{p,2}, u_{p,3}, \ldots, u_{p,U}$, where p is the index number of the physical channel and U is the number of bits in one HS-DSCH subframe (TTI) for one physical channel. For QPSK, U = 960 and for 16QAM U = 1,920. The interleaver is of fixed size: 32 rows and 30 columns. For 16QAM, there are two identical interleavers of the same fixed size. The output bits from the physical channel segmentation are divided two by two between the interleavers: bits $u_{p,k}$ and $u_{p,k+1}$ go to the first interleaver and bits $u_{p,k+2}$ and $u_{p,k+3}$ go to the second interleaver. Bits are collected two by two from the interleavers: bits $v_{p,k}$ and $v_{p,k+1}$ are obtained from the first interleaver and bits $v_{p,k+2}$ and $v_{p,k+3}$ are obtained from the second interleaver, where $k \bmod 4 = 1$.

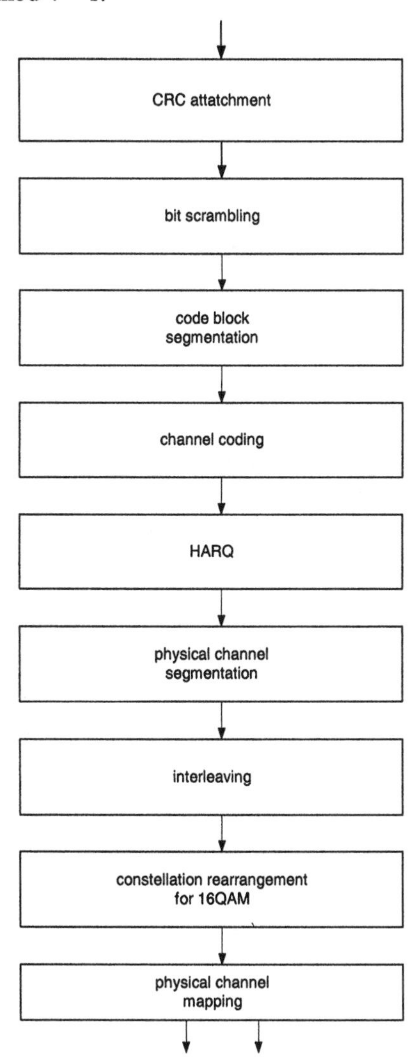

Figure 4.12 Coding chain for HS-DSCH.

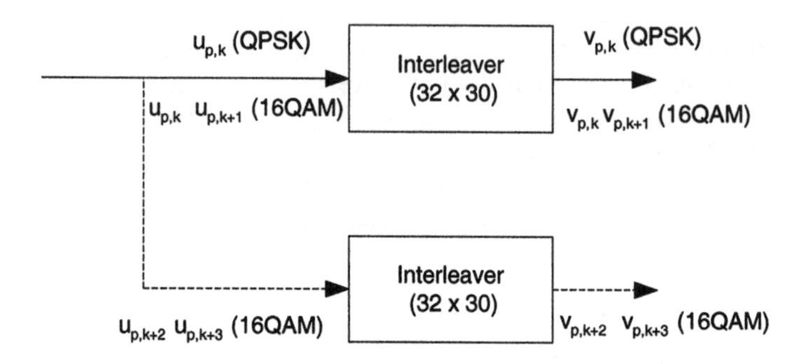

Figure 4.13 The interleaver structure for HS-PDSCH.

4.2.2 HS-SCCH

The HSDPA shared control channel (HS-SCCH) is a downlink physical channel with fixed rate (60 Kbps) and a spreading factor of 128, and it is used to carry the downlink signaling. Figure 4.14 illustrates the subframe structure of the HS-SCCH, which is similar to that of HS-DSCH. It is specified by 3GPP that there can be up to 32 parallel HS-SCCHs in a cell shared by all HSDPA terminals and each terminal can be allocated up to four HS-SCCHs.

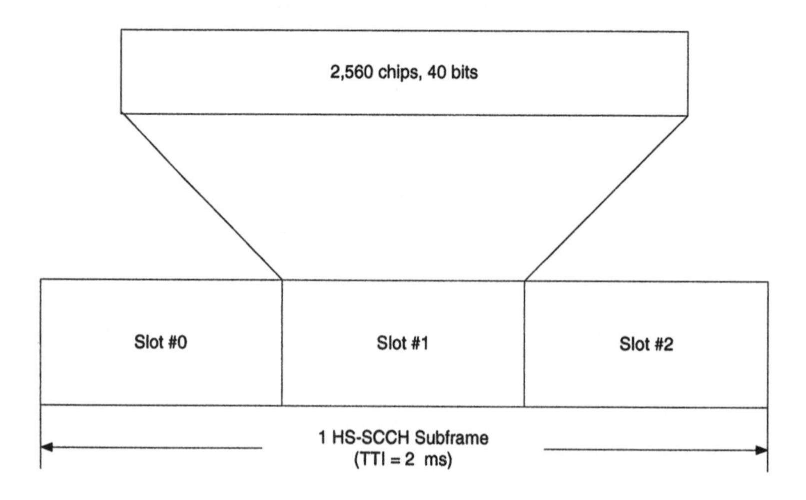

Figure 4.14 Subframe structure of the HS-SCCH.

In each TTI, the HS-SCCH carries HS-DSCH-related downlink signaling for one UE. In the case of HS-DSCH transmission to the same UE in consecutive HS-DSCH TTIs, the same HS-SCCH is used for the corresponding associated downlink signaling. The HS-SCCH power is under the control of the node B. It may follow the power control commands sent by the UE to the node B or any other power control procedures applied by the node B.

The following information is transmitted in the HS-SCCH:

- Channelization code set (7 bits);
- Modulation scheme (1 bit);
- Transport block size (6 bits);
- HARQ process (3 bits);
- Redundancy and constellation version (3 bits);
- New data indicator (1 bit);
- UE identity (16 bits).

4.2.3 Channel Coding for HS-DPCCH

For the HSDPA dedicated physical control channel (HS-DPCCH), the input data to the coding unit includes information on the measured channel quality indication (CQI) and HARQ acknowledgment. The general coding flow is shown in Figure 4.15. It is done in parallel for the HARQ-ACK and CQI, as the flows are not directly multiplexed but are transmitted at different times.

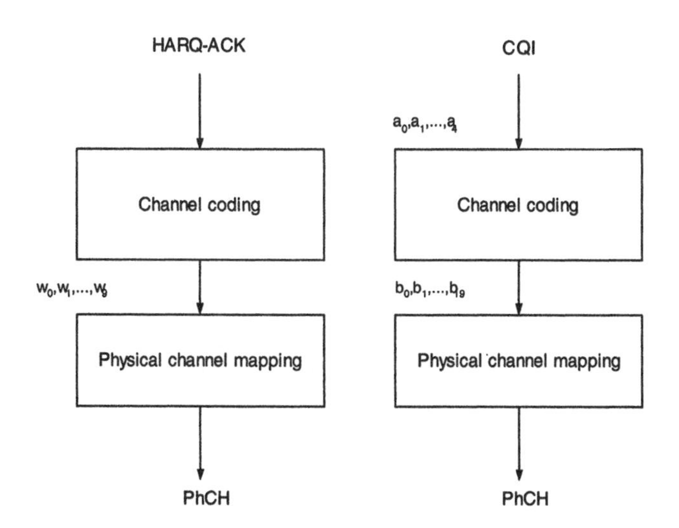

Figure 4.15 Coding for HS-DPCCH.

4.2.3.1 Channel coding for HS-DPCCH HARQ ACK

The HARQ acknowledgment message to be transmitted is coded to 10 bits as shown in Table 4.2. The output is denoted w_0, w_1,…,w_9 and it is seen that 10 ones are used for ACK and 10 zeros are used for NAK.

Table 4.2
Channel Coding of HARQ-ACK

HARQ-ACK message to be transmitted	w_0	w_1	w_2	w_3	w_4	w_5	w_6	w_7	w_8	w_9
ACK	1	1	1	1	1	1	1	1	1	1
NAK	0	0	0	0	0	0	0	0	0	0

4.2.3.2 Channel Coding for HS-DPCCH Channel Quality Information

The channel quality information is coded using a (20,5) code. The code words of the (20,5) code are a linear combination of the five basis sequences denoted $M_{i,n}$ as defined in Table 4.3.

The CQI values [0 .. 30], as explained in Section 4.6.1, are converted from decimal to binary to be mapped to channel quality information bits (0 0 0 0 0) to (0 1 1 1 1) respectively. The channel quality information bits are a_0 , a_1 , a_2 , a_3 , a_4, where a_0 is the least significant bit (LSB) and a_4 is the most significant bit (MSB). The output code word bits b_i are given by:

$$b_i = \sum_{n=0}^{4} (a_n M_{i,n}) \bmod 2 \tag{4.4}$$

where $i = 0, \ldots, 19$.

4.3 MAC-HS

The techniques employed in HSDPA are aimed at facilitating the rapid adaptation of transmission parameters to match channel conditions. Since the node B learns the radio condition change first, the corresponding functions, such as adaptive modulation and coding and fast scheduling, are placed in node B. In contrast, in the Release 99 UTRAN architecture, the scheduling and transport-format selections are performed in the radio network controller (RNC). For HSDPA, it is advantageous to move parts of the functionality from RNC to node B, thus forming a new Node B entity, MAC-hs (see Figure 4.16). The MAC-hs is responsible for handling scheduling, HARQ, and transmit format (TF) selection. The extended features of node B are mainly to support high rate transmission of packet data. Apparently, some upgrading is needed in the node B to enable the MAC-hs functionalities. The consensus among the 3G network vendors is to implement MAC-hs in the channel coding card.

Table 4.3

Basis Sequences for (20,5) Code

i	$M_{i,0}$	$M_{i,1}$	$M_{i,2}$	$M_{i,3}$	$M_{i,4}$
0	1	0	0	0	1
1	0	1	0	0	1
2	1	1	0	0	1
3	0	0	1	0	1
4	1	0	1	0	1
5	0	1	1	0	1
6	1	1	1	0	1
7	0	0	0	1	1
8	1	0	0	1	1
9	0	1	0	1	1
10	1	1	0	1	1
11	0	0	1	1	1
12	1	0	1	1	1
13	0	1	1	1	1
14	1	1	1	1	1
15	0	0	0	0	1
16	0	0	0	0	1
17	0	0	0	0	1
18	0	0	0	0	1
19	0	0	0	0	1

The functional entities included in MAC-hs are shown in Figure 4.16. There is one MAC-hs entity in the UTRAN for each cell supporting HS-DSCH. The MAC-hs is responsible for handling the data transmitted on the HS-DSCH. Furthermore, it is responsible for managing the physical resources allocated to HSDPA. MAC-hs receives configuration parameters from the higher layers. The MAC-hs is comprised of four different functional entities:

- *Flow control*. This is the companion flow control function to the flow control function for existing dedicated, common, and shared channels in RNC. This function is employed to limit layer 2 signaling latency and reduce discarded and retransmitted data as a result of HS-DSCH congestion.
- *Scheduling/Priority handling*. This function manages HS-DSCH resources between HARQ entities and data flows according to their priority. There is one priority queue for each MAC-d protocol data unit (PDU) in the MAC-hs. Based on status reports from associated uplink signaling in HS-DCCH, either new transmission or retransmission is determined. A new transmission can be initiated instead of a pending retransmission at any time to support the priority handling.
- *HARQ*. One HARQ entity handles the hybrid ARQ functionality for one user. One HARQ entity is capable of supporting up to eight HARQ processes of stop-and-wait HARQ protocols. There is one HARQ process per HS-DSCH per TTI.

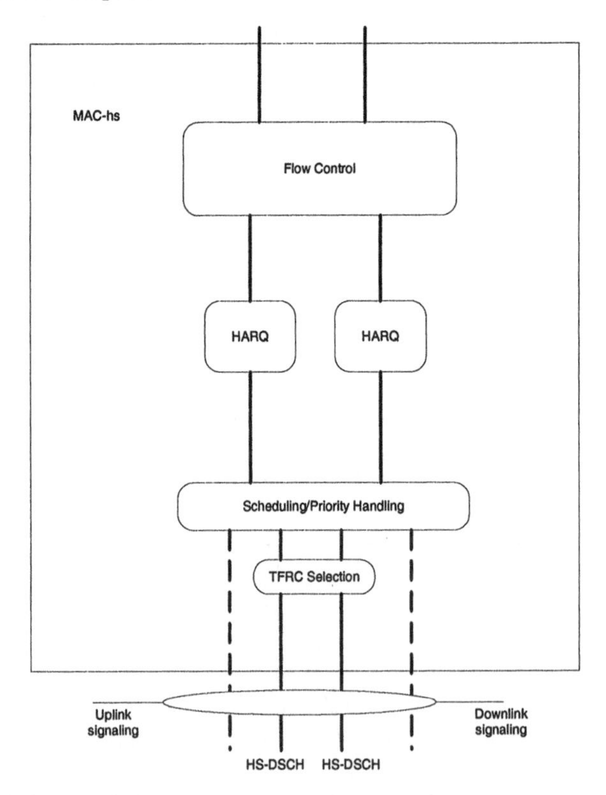

Figure 4.16 Functional entities in MAC-hs.

- *TFRC selection.* This is to select an appropriate transport format and resource combination (TFRC) for the data to be transmitted on HS-DSCH.

The associated signaling shown in Figure 4.16 illustrates the exchange of information between layer 1 and layer 2. In the following, the functions of two major functional entities in MAC-hs, the scheduler and the HARQ unit, are explained further.

4.3.1 Scheduler

As described in Section 4.1.3, the scheduler is one of the most important functional entities in determining the QoS and data rate of HSDPA services, as it controls when and how to transmit data streams dedicated at each terminal. For each terminal, the information available to the scheduler includes the estimate of channel quality (CQI) received on the HS-DPCCH, the knowledge of priority queues, and the HARQ processes and terminal capability. Based on the information, the scheduler performs the following functions:

- Schedules all HSDPA users within a cell.
- Services priority queues.
- Schedules MAC-hs PDUs based on information from the HS-DSCH frame protocol. One terminal may be associated with one or more MAC-d flows. Each MAC-d flow contains HS-DSCH MAC-d PDUs for one or more priority queues.
- Determines the HARQ entity and the queue to be serviced.
- Indicates the queue ID and the transmit sequence number (TSN) to the HARQ entity for each MAC-hs PDU to be transmitted.
- Schedules new transmissions and retransmissions.
- Based on the status reports from HARQ processes, the scheduler determines if either a new transmission or a retransmission should be made. A new transmission can, however, be initiated on a HARQ process at any time.
- Determines a suitable redundancy version and modulation scheme for each transmitted and retransmitted MAC-hs PDU and indicates the redundancy version to lower layers.

4.3.2 HARQ Unit

The HARQ unit is responsible for handling the HARQ functionalities of all mobile terminals. There is one HARQ functional entity per mobile terminal in UTRAN. Each functional entity can manage up to eight parallel stop-and-wait HARQ processes. As the input, the HARQ entity receives the acknowledgment (ACK/NAK) from the mobile terminal. The HARQ entity sets the queue ID in transmitted MAC-hs PDUs based on the identity of the queue being serviced. The HARQ entity sets the transmission sequence number (TSN) in transmitted MAC-hs PDUs. The TSN is set to value 0 for the first MAC-hs PDU transmitted for one HS-DSCH and queue ID and it is increased by one for each subsequent transmitted MAC-hs PDU. The HARQ entity determines a suitable HARQ process to service the MAC-hs PDU and sets the HARQ process identifier accordingly.

The HARQ process sets the new data indicator in the transmitted MAC-hs PDUs. It sets the new data indicator to value "0" for the first MAC-hs PDU transmitted by a HARQ process and then increases the new data indicator with one for each transmitted MAC-hs PDU containing new data. The HARQ processes received status messages. UTRAN delivers received status messages to the scheduler.

4.3.3 Interworking within MAC-hs

To summarize, the MAC-hs needs to perform the following tasks when dealing with HSDPA traffic:

- Decode the higher-layer information regarding UE capability and required QoS for initial connection.
- Decode the ACK/NAK and CQI transmitted in the uplink and check the power level of the downlink dedicated physical channel.
- Make a scheduling decision on which terminal is due to receive data among the terminals having data in the transmission buffer.
- Set the transport format combination indicator (TFCI) and HARQ parameters in the downlink shared control channel HS-SCCH.

The interworking of different entities within MAC-hs is shown in Figure 4.17.

4.4 RADIO RESOURCE MANAGEMENT

Radio resource management is a complicated but very important task in UTRAN networks. Its functions include admission control, spreading code allocation, radio bearer control, power control, handover control, load control, and congestion control. For voice only services, the radio resources management is mainly aimed at managing the interference level. This is because the system capacity is directly dependent on the interference caused by different signals to each other and minimizing interference can increase the quality of services and maximize the system capacity. When HSDPA is supported, however, the radio resource management has the additional task of managing packet connections and bursty transmissions. Depending on the specific applications, data services have a variety of data rate requirement, frame error rate requirement, jitter, and delay requirement. This demands a closer collaboration between RNCs and node Bs. As discussed in Section 4.1, the scheduler in node B plays an important role in controlling the data rate and quality of services. In the meantime, the admission control in the controlling RNC must make sure that users admitted to the HSDPA service are allocated with adequate spreading codes and power and there are enough resources including codes, power and time slots to meet the required data rate and quality of services of the accessing user.

Code allocation for HSDPA is done in two stages. First, the controlling RNC (CRNC) assigns a set of codes to each cell for HS-PDSCH and HS-SCCH that share a common scrambling code [14, 15]. Up to 15 HS-PDSCH codes with a spreading factor of 16 and up to 32 HS-SCCH with a spreading factor of 128 can be assigned for each cell. As both the HS-PDSCH and the HS-SCCH share a common code tree, it is not possible to assign 15 HS-PDSCH codes and 32 HS-SCCH codes to a cell simultaneously. Second, the scheduler dynamically allocates an appropriate number of codes for each terminal and for each transmit time interval (TTI). If the HS-PDSCH and non-HS-PDSCH channels are operating on the same carrier, the CRNC manages the number of spreading codes used by HS-PDSCH dynamically. If the number of codes allocated to HS-DSCH is not enough, it is logical to allocate more, provided that there is a spare code resource. This can be done either upon a request by the node B, or by a decision made by the controlling RNC upon receiving a code utilization measurement report from the node B that indicates the actual number of codes that have been used and the frequency of such usage [16].

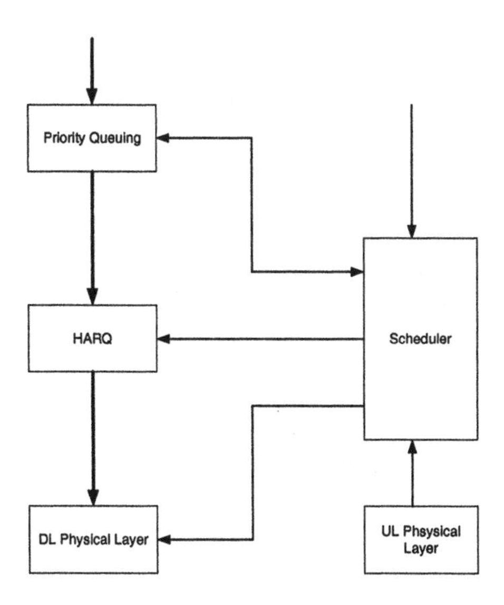

Figure 4.17 Interworking of MAC-hs entities.

For the HSDPA physical channel, HS-PDSCH, the open power control is used with a target set by the controlling RNC. In other words, the controlling RNC may set a power limit for the total power of all the HS-PDSCH and HS-SCCH channels in a cell and the node B in turn performs the open loop power control for HS-DSCH based on the channel quality measurements reported by the terminals or the power control parameters of the associated DL-DPCH.

It should be mentioned that, in order to support HSDPA, new measurements for HSDPA are needed in node B [17]. This is because the controlling RNC needs to know the power distribution between HSDPA and non-HSDPA channels in order to perform admission control. Since scheduling is done in node B, the controlling RNC does not have the information about the real-time usage of the node B resources. Such information is necessary to enable the controlling RNC to make the decision about the admission of a new call and to meet the requested quality of services. For Release 99 systems, the controlling RNC has the information on the total downlink Tx power. For Release 5 systems, the non-HSDPA power measurement for dedicated, common, and shared channels will be reported by node B. By subtracting these two measurements, the power used by HSDPA channels can be obtained in the controlling RNC.

In the current UTRAN Release 5 specification, the information on the resource usage in node B with respect to each priority of streaming class is missing. It would be beneficial to obtain the information regarding the amount of resources required to meet the QoS requirement of all connections per given priority class. Such information should be based on feasible measurements carried out within node B. Guaranteed bit rate (GBR) and transfer delay are two critical parameters for streaming class. The actual bit rate and the power used to provide it may vary with time, as the actual peak data rate may be higher than the guaranteed one or even lower if the channel condition of the terminal in question is poor.

In fact, based on the measurement of the provided bit rate, the provided power, and the knowledge of GBR per particular connection, one can calculate the power required for the guaranteed bit rate. Furthermore, one can sum these requirements over all users of the given priority class and signal this value from node B to the controlling RNC. In essence, it is preferable that in node B common measurement "Power for GBR" is done per priority class averaged over a defined period [18].

4.5 MOBILITY PROCEDURES

Once a terminal is in the so-called CELL_DCH state when dedicated channels have been set up, it can be allocated with one or more HS-PDSCH(s), thus allowing it to receive data on the HS-DSCH. For dedicated channels, it is advantageous to employ the so-called soft handover technique, which is to transmit the same data from a number of node Bs simultaneously to the terminal, as this provides diversity gain. Owing to the nature of packet transmission, however, synchronized transmission of the same packets from different cells is very difficult to achieve, so only hard handover is employed for HS-PDSCH. This is referred to HS-DSCH cell change, and the terminal can have only one serving HS-DSCH cell at a time [14]. A *serving HS-DSCH cell change* message facilitates the transfer of the role of serving HS-DSCH radio link from one belonging to the source HS-DSCH cell to another belonging to the target HS-DSCH cell.

In theory, the serving HS-DSCH cell change can be decided either by the mobile terminal or by the network. In UTRAN Release 5, however, only network-controlled serving HS-DSCH cell changes are supported and the decision can be based on UE measurement reports and other information available to the RNC. A network-controlled HS-DSCH cell change is performed based on the existing handover procedures in CELL_DCH state.

Since the HSDPA radio channel is associated with dedicated physical channels in both the downlink and uplink, there are two possible scenarios in changing a serving HS-DSCH cell: (1) only changing the serving HS-DSCH cell and keeping the dedicated physical channel configuration and the active set for handover intact; or (2) changing the serving HS-DSCH cell in connection with an establishment, release, and/or reconfiguration of dedicated physical channels and the active set.

Although an unsynchronized serving HS-DSCH cell change is permissible, a synchronized one is obviously preferable for ease of traffic management. In that case, the start and stop of the HS-DSCH transmission and reception are performed at a given time. This is convenient especially when an intranode B serving HS-DSCH cell change is performed, in which case both the source and target HS-DSCH cells are controlled by the same node B and the change happens between either frequencies or sectors.

If an internode B serving HS-DSCH cell change is needed, the *serving HS-DSCH Node B relocation* procedure needs to be performed in the UTRAN. During the serving HS-DSCH node B relocation process, the HARQ entities located in the source HS-DSCH node B belonging to the specific mobile terminal are deleted and new HARQ entities in the target HS-DSCH node B are established. In this scenario, different controlling RNCs may control the source and target HS-DSCH node Bs, respectively.

4.5.1 Intranode B Serving HS-DSCH Cell Change

Figure 4.18 illustrates an intranode B serving HS-DSCH cell change while keeping the dedicated physical channel configuration and the active set, using the physical channel reconfiguration procedures. The transition from source to target HS-DSCH cells is performed in a synchronized fashion, that is, at a given activation time. For clarity, only the layers directly involved in the process are shown and the sequence of the events starts from the top and finishes at the bottom.

In this scenario, the terminal transmits a *measurement report* message containing intrafrequency measurement triggered by the event *change of best cell.* When the decision to perform handover is made at the serving RNC (SRNC), the node B is prepared for the serving HS-DSCH cell change at an activation time indicated by *CPHY-RL-Commit-REQ primitive.* The serving RNC then sends a *physical channel reconfiguration* message, which indicates the target HS-DSCH cell and the activation time to the UE. Since the same node B controls both the source and target HS-DSCH cells, it is not necessary to reset the MAC-hs entities. Once the terminal has completed the serving HS-DSCH cell change, it transmits a *physical channel reconfiguration complete* message to the network.

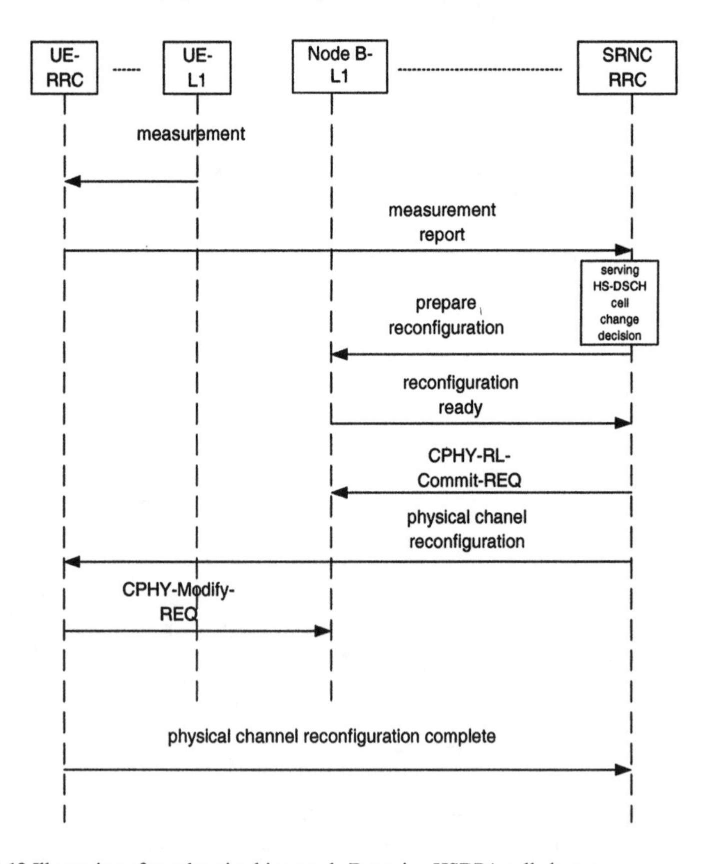

Figure 4.18 Illustration of synchronized intranode B serving HSDPA cell change.

It should be pointed out that, in this particular case, it is assumed that HS-DSCH transport channel and radio bearer parameters do not change. If transport channel or radio bearer parameters are changed, the serving HS-DSCH cell change would need to be executed by a transport channel reconfiguration procedure or a radio bearer reconfiguration procedure, respectively.

4.5.2 Internode B Serving HS-DSCH Cell Change

For terminals on the move, what happens more often than the intra-node B serving HS-DSCH cell change is the so-called internode B serving HS-DSCH cell change. For synchronized case, the reconfiguration is performed in two steps within UTRAN.

To begin with, the terminal transmits a *measurement report* message containing measurement triggered by the event *change of best cell*. The serving RNC determines the need for hard handover based on received measurement report and/or load control algorithms. As the first step, the serving RNC establishes a new radio link in the target node B. After this, the target Node B starts transmission and reception on dedicated channels. In the second step, this newly created radio link is prepared for a synchronized reconfiguration to be executed at a given activation time indicated in the CPHY-RL-Commit-REQ primitive, at which the transmission of HS-DSCH will be started in the target HS-DSCH node B and stopped in the source HS-DSCH node B.

The serving RNC then sends a *transport channel reconfiguration* message on the old configuration. This message indicates the configuration after handover, both for DCH and HS-DSCH. The transport channel reconfiguration message includes a flag indicating that the MAC-hs entity in the terminal should be reset. The message also includes an update of transport channel-related parameters for the HS-DSCH in the target HS-DSCH cell.

After physical synchronization is established, the terminal sends a *transport channel reconfiguration complete* message. The serving RNC then terminates reception and transmission on the old radio link for dedicated channels and releases all resources allocated to the UE. The process of internode B handover for HS-DSCH is shown in Figure 4.19.

It should be noted that in the case of internode B handover, the radio link control (RLC) for transmission/reception on HS-DSCH may be stopped at both the UTRAN and the terminal sides prior to reconfiguration and continued when the reconfiguration is completed, which could result in data loss during the handover period. Furthermore, the *transport channel reconfiguration* message indicates to the terminal that the MAC-hs entity should be reset and a status report for each RLC entity associated with the HS-DSCH should be generated. However, a reset of the MAC-hs entity in the terminal does not require flushing the reordering buffers but delivering the content to higher layers.

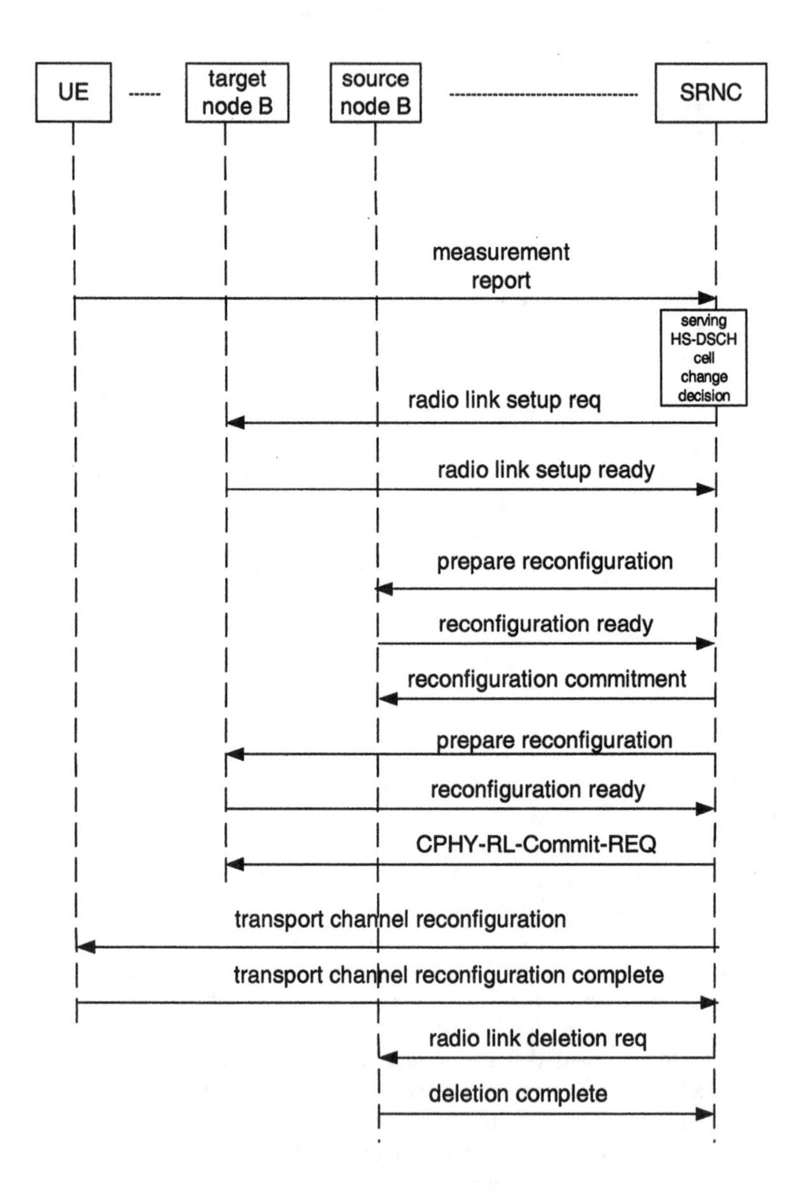

Figure 4.19 Illustration of synchronized internode B serving HSDPA cell change.

4.6 HSDPA IMPACT ON MOBILE TERMINALS

Although this book is mainly focused on the radio access network, a discussion on the HSDPA impact on mobile terminals is useful in understanding both the operation of HSDPA and the complexity of the technology. Therefore, three topics are discussed briefly in this section, namely, the operation of mobile terminals, the buffering complexity, and the signal processing required to support HSDPA.

4.6.1 Terminal Operation

Before setting up the HS-DSCH connection, the following physical layer parameters are signaled from node B to the mobile terminal from higher layers in order for the latter to obtain the control information and to send feedback information in the correct format:

- HS-SCCH set to be monitored;
- Repetition factor of ACK/NACK: *N_acknak_transmit*;
- Channel quality indicator (CQI) feedback cycle k;
- Repetition factor of CQI: *N_cqi_transmit*;
- Measurement power offset Γ.

If a terminal cannot detect the control information intended for it on any of the HS-SCCHs in the HS-SCCH set in the previous subframe, it should monitor all HS-SCCHs in the HS-SCCH set. If the terminal has detected the control information intended for it in the previous subframe, it is sufficient to only monitor the same HS-SCCH used in the previous subframe. Once the terminal has found out that one of the monitored HS-SCCHs carries the control information intended for it, the terminal should start receiving the HS-PDSCHs indicated by this control information. The transport block size information is derived from the signaled transport format combination indicator (TFCI).

After decoding the HS-PDSCH data, the terminal transmits a HARQ ACK or NAK as determined by the MAC-hs based on the CRC check. The terminal repeats the transmission of the ACK/NAK information over *N_acknak_transmit* consecutive HS-DPCCH sub-frames in the slots allocated to the HARQ-ACK. When *N_ acknak_transmit* is greater than 1, the terminal should not attempt to receive or decode transport blocks from the HS-PDSCH in the $n+1$ to $n + (N\_acknack\ \_transmit - 1)$ HS-DSCH subframes, where n is the number of the last HS-DSCH sub-frame in which a transport block has been received. If the control information is not detected on any of the HS-SCCHs in the HS-SCCH set, neither ACK nor NAK is transmitted in the corresponding subframe.

One major impact of HSDPA on the terminal is that the latter must be equipped with an HARQ entity to support the HARQ protocol. There is one HARQ entity in each terminal which processes the HARQ process identifiers in the received MAC-hs PDUs on HS-DSCH. Each received MAC-hs PDU is allocated to the HARQ process indicated by the HARQ process identifier of the MAC-hs PDU. Up to eight parallel HARQ processes can be used in a terminal to support the HARQ protocol. The actual number of HARQ processes is configured by higher layers. The HARQ process processes the new data indicator indicated by lower layers for each received MAC-hs PDU. The terminal takes the following actions according to different scenarios.

- If the new data indicator has been incremented compared to the value in the previous received transmission in this HARQ process or this is the first received transmission in the HARQ process, the terminal replaces the old data stored in the soft buffer for this HARQ process with the received data. This is the case when the previous transmission has been successfully received.
- If the new data indicator is identical to the value used in the previous received transmission in the HARQ process, the terminal combines the received data with the data currently in the soft buffer for this HARQ process. This is the case when the previous detection was unsuccessful and Chase combining is needed.
- If the data in the soft buffer has been successfully decoded and no error has been detected, the terminal delivers the decoded MAC-hs PDU to the reordering entity and generates a positive acknowledgment (ACK) of the data in this HARQ process. Otherwise, the terminal generates a negative acknowledgment (NAK) of the data in this HARQ process.

The terminal schedules the generated positive or negative acknowledgment for transmission. The time of transmission relative to the reception of data in a HARQ process is configured by higher layers. The HARQ processes the queue ID in the received MAC-hs PDUs. The terminal arranges the received MAC-hs PDUs in queues based on the queue ID.

Another important functionality for the terminal is to provide feedback on the channel conditions by sending channel quality indicator (CQI) on the HS-DPCCH. Based on an unrestricted observation interval, the terminal reports the highest tabulated CQI value. This CQI value suggests the transport block size, the number of HS-PDSCH codes and the modulation scheme that can be used by the node B in a single HS-DSCH subframe. The guaranteed transport block error probability is 0.1. The CQI value is determined by both the terminal capability and the measured power in the common pilot channel (CPICH). There are 30 different CQI values and each corresponds to a different combination of the modulation scheme, the number of HS-PDSCH codes, and the transport block size.

For the frequency division duplex (FDD) mode, UTRAN Release 5 defines 12 UE categories as shown in Table 4.4. It is shown that UEs of categories 11 and 12 support QPSK only, whereas UEs in other categories support both QPSK and 16QAM. It should be noted in Table 4.4 that not all UEs support 15 HS-PDSCH codes. In fact, the majority of them only support five HS-PDSCH codes.

Table 4.4
Categories of FDD Terminals

HS-DSCH Category	*Maximum Number of HS-DSCH Codes Received*	*Minimum Inter-TTI Interval*	*Maximum Number of Bits of an HS-DSCH Transport Block Received within an HS-DSCH TTI*
Category 1	5	3	7,300
Category 2	5	3	7,300
Category 3	5	2	7,300
Category 4	5	2	7,300
Category 5	5	1	7,300
Category 6	5	1	7,300
Category 7	10	1	14,600
Category 8	10	1	14,600
Category 9	15	1	20,432
Category 10	15	1	28,776
Category 11	5	2	3,650
Category 12	5	1	3,650

For the purpose of CQI reporting, the terminal assumes a total received HS-PDSCH power of the following

$$P_{HS-PDSCH} = P_{CPICH} + \Gamma + \Delta \tag{4.5}$$

and it also assumes that the total received power is evenly distributed among the HS-PDSCH codes of the reported CQI value. The measurement power offset Γ is signaled by higher layers and the reference power adjustment Δ is defined in the standards. As an example, Table 4.5 shows a CQI mapping table for terminals of categories 1 to 6.

If the higher layer signaling informs the terminal that, for the radio link in the serving HS-DSCH cell, it may use a secondary common pilot channel (S-CPICH) as a phase reference and the P-CPICH is not a valid phase reference, P_{CPICH} in (4.6) should be interpreted as the received power of the S-CPICH used by the terminal; otherwise, P_{CPICH} is the received power of the P-CPICH. If transmit diversity is used for the radio link from the serving HS-DSCH cell, P_{CPICH} in (4.6) denotes the received power of the combined CPICH from both diversity antennas, otherwise it denotes the power from the nondiversity antenna.

Table 4.5
CQI Mapping for Terminals of Category 1 to 6

CQI Value	*Number of HS-PDSCHs*	*Modulation Scheme*	*Δ*
0	Out of range		
1	1	QPSK	0
2	1	QPSK	0
3	1	QPSK	0
4	1	QPSK	0
5	1	QPSK	0
6	1	QPSK	0
7	2	QPSK	0
8	2	QPSK	0
9	2	QPSK	0
10	3	QPSK	0
11	3	QPSK	0
12	3	QPSK	0
13	4	QPSK	0
14	4	QPSK	0
15	5	QPSK	0
16	5	16-QAM	0
17	5	16-QAM	0
18	5	16-QAM	0
19	5	16-QAM	0
20	5	16-QAM	0
21	5	16-QAM	0
22	5	16-QAM	0
23	5	16-QAM	-1
24	5	16-QAM	-2
25	5	16-QAM	-3
26	5	16-QAM	-4

Table 4.5 (Continued)

27	5	16-QAM	-5
28	5	16-QAM	-6
29	5	16-QAM	-7
30	5	16-QAM	-8

4.6.2 Buffering Complexity

The requirement of HARQ on mobile terminals is to buffer HSDPA data that have not been received correctly and combine them with the retransmitted ones for further decoding attempts. The actual method of soft combining depends on the HARQ combining scheme employed. In the Chase combining scheme, the receiver always combines the full retransmission of the failed data in the HSDPA transmit time interval (TTI) and this means that the amount of data in the receiver buffer remains the same. In the incremental redundancy schemes, every consecutive transmission brings in new information to the receiver, so the amount of data to be buffered increases with consecutive retransmissions. Regardless of the HARQ combining scheme, soft combining is done on physical layer before the decoding stage of FEC. Prior to decoding, these symbols are soft-valued, that is, each symbol is represented by two or more bits.

For dual channel stop-and-wait HARQ the buffer size estimation is straightforward, as no new data are transmitted on any subchannel before the previous packet is acknowledged. The receiver needs to buffer one HSDPA TTI for each subchannel. The next transmission is either a new packet or a retransmission of the incorrectly received packet. In either case, the maximum buffering need is two HSDPA TTIs. This receiver buffering complexity estimate can be easily extended to an n-channel stop-and-wait protocol. At maximum, data in n HSDPA TTIs need to be buffered at any given time.

4.6.3 Signal Processing Required

HSDPA introduces some new signal processing tasks to the mobile terminals. The major ones include the following:

- Decode TFCI, HSDPA TTI sequence information.

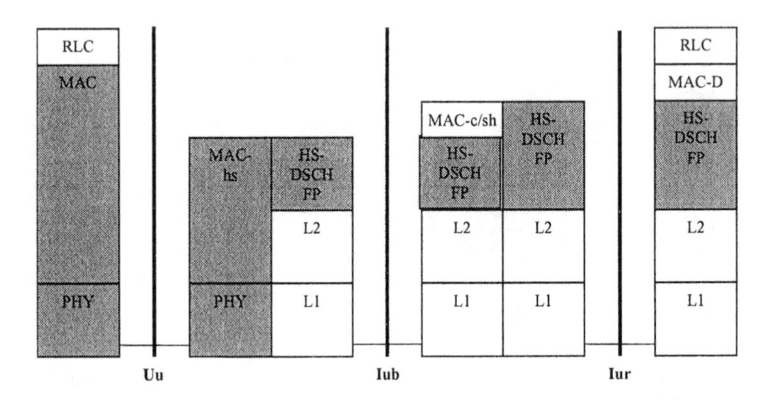

Figure 4.20 UTRAN protocol architecture supporting HSDPA. Shaded areas are the protocol layers affected by HSDPA.

- Soft combine retransmitted HSDPA TTI in the receiver buffer for the correct subchannel.
- Decode HSDPA TTI for the subchannel.
- Check CRC to decide whether the HSDPA TTI has been received correctly.
- Generate ACK/NAK signaling for the uplink.

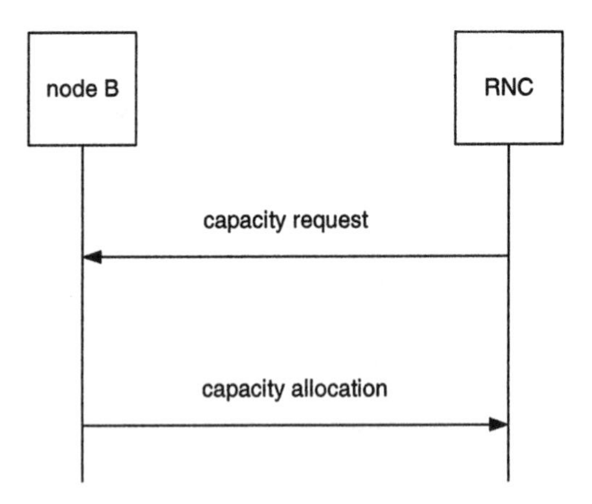

Figure 4.21 HS-DSCH FP control procedures.

4.7 HSDPA PROTOCOL ARCHITECTURE

Figure 4.20 shows the overall UTRAN protocol architecture to support HSDPA. It is seen that, in addition to MAC-hs in node B, a HS-DSCH frame protocol (FP) is also introduced. The HS-DSCH FP is terminated at the serving RNC. Dedicated MAC (MAC-d) entity in the RNC transfers MAC-d PDUs to the MAC-hs in the node B using the services of HS-DSCH FP. The HS-DSCH FP facilitates multiplexing of data streams from different terminals onto the same data frame and allows multiple terminals and multiple MAD-d flows to share the same transport bearer. Since the RLC and MAC-d layers are unchanged from Release 99 architecture, the HW impact of HSDPA on RNC HW is limited.

Figure 4.21 shows the HS-DSCH FP control procedures. The HS-DSCH FP provides flow control mechanism by using two control messages on the Iub interface [19]. The first is HS-DSCH capacity request message sent from RNC. It is used to request HS-DSCH capacity from node B by indicating the user buffer size of RNC for a given priority level. The message is normally triggered by an event, such as data arrival. The second one is HS-DSCH capacity allocation message, which is generated within node B and sent to RNC. In the message, the *HS-DSCH credit information element* (IE) is used to indicate the number of MAC-d PDUs that an RNC may transmit during one HS-DSCH interval. This message provides means to control user data flow from RNC to node B. It can be sent anytime to modify the capacity of user plane data flow. If the *HS-DSCH credit* IE has a value of 0, it signifies that there are no resources allocated for transmission, so the RNC should stop transmission. If the *HS-DSCH credit* IE has a value of 2,047, it signifies unlimited capacity for the transmission of PDUs.

Once the RNC has been allocated capacity by the node B via the HS-DSCH capacity allocation control frame or via the HS-DSCH initial capacity allocation and the RNC has data waiting to be sent, the HS-DSCH data frame is used to transfer the data. If the RNC has been granted capacity by the node B via the HS-DSCH initial capacity allocation, this capacity is valid for only the first HS-DSCH data frame transmission. When data is waiting to be transferred and a capacity allocation message is received, a data frame will be transmitted immediately according to the allocation received. Multiple MAC-d PDUs of the same length and the same priority level can be transmitted in one MAC-d flow in the same HS-DSCH data frame.

The HS-DSCH data frame includes a user buffer IE to indicate the amount of data pending for the respective MAC-d flow for the indicated priority level. Within one priority level and size, the MAC-d PDUs are transmitted by the node B on the Uu interface in the same order as they are received from the RNC.

4.8 HSDPA DEPLOYMENT

By the time the HSDPA technology is introduced, most mobile operators in western Europe will have installed at least part of the UTRAN network. Therefore, it is very important for the technology to be introduced in such a manner that maximum infrastructure reuse is realized. Although this applies to all parts of the network, it is the base station that will be affected the most by HSDPA.

Owing to the large number of radio base station sites, on-site updates can cause major logistics problems and increase the operation and maintenance cost of the mobile operators. To facilitate future migration, base stations are usually designed to accommodate years of feature upgrades by means of remote software loading. Since the UTRAN standards for HSDPA only became stable in 2003, however, base stations that were installed in the early roll-out may need on-site hardware upgrades. Fortunately, the modular radio base station architecture used by many vendors is designed to accommodate such changes so the hardware change will be limited. To enable the UTRAN network evolution into HSDPA, two alternative paths can be taken. The first path is node B upgrade, wherein only the channel coding card replacement and software upgrade in the existing node Bs are required. The second path is for green-fielders and late UTRAN adopters, wherein HSDPA-enabled node Bs will be provided or HSDPA is only part of the optional feature set supported by software. In any case, it is expected that not all node Bs will be able to support HSDPA when the service is first introduced. Therefore, as shown in Figure 4.22, the RNCs will have the responsibility of managing mixed types of node Bs. One important impact of HSDPA on RNC is that the Iub must be able to handle much higher throughput, which could cause some limited hardware replacement in the transport network if the capacity has not been provisioned.

There are a variety of options to deploy HSDPA. Operators can choose to place HSDPA related-channels on the same carrier used by Release 99 services to improve the efficiency of packet based data services and to increase cell capacity. In this case, Release 99 and HSDPA-enabled terminals will be able to camp simultaneously on the same carrier. This permits smooth introduction of HSDPA and operators' investment can be kept in proportion to traffic demand. Another option is to place HSDPA services on a separate carrier. Effectively, this will lead to an overlay of HSDPA cells and Release 99 cells. The advantage of the second option is that the two types of cells can be easily optimized to support data and voice traffic, respectively. In fact, the second approach has actually been taken by CDMA2000 1xEV-DO. However, the disadvantage of the second option is that more investment is needed in the RF part and the terminals can become more expensive.

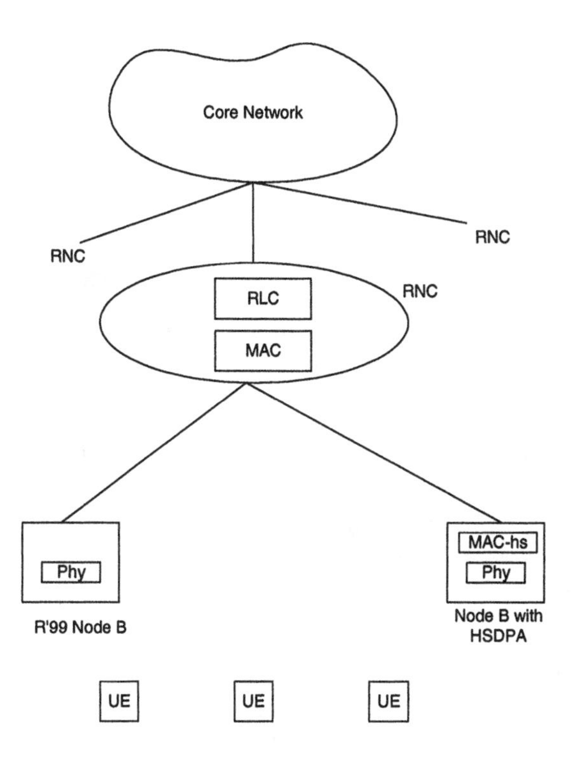

Figure 4.22 The mixed architecture for HSDPA deployment.

Initially, the HSDPA services will probably only be offered in part of the network, such as airports and city centers. When a user leaves the area with HSDPA coverage, an ongoing call can be transparently switched from the HS-DSCH to any of the channels already existing in Release 99 systems using the channel switching and handover mechanisms provided by UTRAN. For the end user, the only noticeable effect of leaving the HSDPA area will be reduced data rates. Thus, mobile operators need not upgrade all cells in the network simultaneously. Instead, they can gradually enhance the network with HSDPA as the demand for capacity increases. Furthermore, the fraction of resources in a cell used by HSDPA can be dynamically configured to match long-term traffic variations. Indoor environments represent one deployment scenario in which the high data rates of HSDPA can be exploited to their full extent. A cost-effective way of providing indoor coverage can be using distributed antennas with remote RF heads, as discussed in Chapter 2.

With regard to terminals that support HSDPA, mobile operators have the flexibility of supporting different capability classes, as shown in Section 4.6. A terminal with limited memory and QPSK modulation will not support the highest data rates, but its cost will be relatively low. On the other hand, a high-end terminal that supports all modulation and coding schemes can take full advantage of the HSDPA technology. This implies that users who expect good propagation conditions can fully exploit the advantage of HSDPA using a high-end terminal without imposing extra costs on users who settle for medium data rates and less expensive terminals. An interesting aspect of best-effort systems is that user terminals with better receiver performance can enjoy a higher data rate. Thus, there is an incentive for providing improved receiver algorithms by terminal vendors and for users to purchase more expensive terminals.

It should be pointed out that the user throughput and system throughput achievable in a real system depend on both the implementation parameters and terrain conditions. To achieve the high data rate while maintaining coverage, the HS-PDSCH will use up a significant amount of the code and power resource. When provided on the same carrier, HSDPA services will interfere with non-HSDPA services. Therefore, the maximum data rate provided by HSDPA may differ from the theoretical predictions. For these reasons, it is envisaged that some operators may choose to use dedicated carriers for HSDPA services. This will not only ease the interference problem but also realize the full potential of HSDPA.

4.9 CONCLUDING REMARKS

HSDPA represents a great step in the evolution of mobile radio access networks. First, it gives certain access control and radio resource management power to the base stations to exploit the diversity and dynamics of the local radio environment, thus leading to more efficient utilization of the radio channel and the spectrum. Second, by placing flow control, scheduling, and HARQ functionalities at the edge of the radio access network, the processing burden on the centralized RNC is reduced. This also reduces the risk of network breakdown caused by RNC failure. Third, the efficient packet transmission provided by HSDPA paves the way for all IP networks (see Chapter 7).

To improve the performance of HSDPA, further increase data rate, and reduce delay, 3GPP is working to introduce some enhancement technologies [20]. Undoubtedly, this will lead to higher throughput for the end users and further enrich user experience. A good example technique under discussion is called fast adaptive emphasis, which modifies the feedback information for closed-loop transmit diversity so that the maximum diversity gain is obtained for either HSDPA or dedicated channels on a dynamic basis. In addition to technologies being standardized, one could also use various multiple antenna technologies for HSDPA services to increase coverage and reduce interference to other user terminals. One should note, however, that since many multiple antenna technologies are based on the exploitation of the diversity of radio channels, just as with HSDPA, care must be taken to integrate and optimize the whole system so that the net gain achieved can justify the overall cost and complexity [21].

References

[1] S. Parkvall et al, "WCDMA Evolved – High Speed Packet Data Services," *Ericsson Review*, 2/2003, pp. 56 – 65.

[2] 3GPP TS 25.855 v5.0.0, High Speed Downlink Packet Access; Overall UTRAN Description (Release 5), http://www.3gpp.org.

[3] 3GPP TS 25.308, v5.3.0, Technical Specification Group Radio Access Networks; High Speed Downlink Packet Access (HSDPA); Overall Description; Stage 2 (Release 5), http://www.3gpp.org.

[4] 3GPP TS 25.858 v5.0.0, High Speed Downlink Packet Access, Physical Layer Aspects (Release 5), http://www.3gpp.org.

[5] L. Hanzo, W. Webb, and T. Keller, *Single and Multi-Carrier Quadrature Amplitude Modulation*, New York: John Wiley & Sons, 2000.

[6] 3GPP TS 25.213 v5.3.0, Technical Specification Group Radio Access Networks; Spreading and Modulation (Release 5), http://www.3gpp.org.

[7] W. Stallings, *Data and Computer Communications*, Upper Saddle River, NJ: Prentice-Hall, 1997.

[8] S. Lin, D. J. Costello, Jr., and M. J. Miller, "Automatic-Repeat-Request Error Control Schemes," *IEEE Communications Magazine*, Vol. 22, No. 12, December 1984, pp. 5-17.

[9] R. H. Deng, "Hybrid ARQ Schemes Employing Coded Modulation and Sequence Combining," *IEEE Trans. on Communications*, Vol. 42, June 1994, pp. 2239-2245.

[10] 3GPP TR 25.848 v4.0.0, Technical Specification Group Radio Access Networks; Physical Layer Aspects of UTRA High Speed Downlink Packet Access (Release 4), http://www.3gpp.org.

[11] O. Lataoui, et al., "A QOS Management Architecture for Packet Switched 3rd Generation Mobile Systems," *NetWorld+Inerop2000 – Engineers Conference on Broadband Internet Access Technologies Systems & Services*, May 2000.

[12] J. Holtzman, "CDMA Forward Link Waterfilling Power Control," *Proceedings of VTC2000 –* Spring, May 2000, pp. 1663-1667.

[13] 3GPP TS25.212 v5.4.0, Technical Specification Group Radio Access Networks; Multiplexing and Channel Coding (FDD) (Release 5), http://www.3gpp.org.

[14] 3GPP TS25.433, Technical Specification Group Radio Access Networks; UTRAN Iub Interface NBAP Signaling, http://www.3gpp.org.

[15] 3GPP TS 25.877 v5.1.0, Technical Specification Group Radio Access Networks; High Speed Downlink Packet Access: Iub/Iur Protocol Aspects (Release 5), http://www.3gpp.org.

[16] Nokia, HS-PDSCH Code Utilization Measurement, R3-030571, 3GPP TSG RAN WG3 Meeting #36, http://www.3gpp.org.

[17] NEC and Telecom Modus, Common Measurements for HS-DSCH Redimensioning and CAC for HSDPA, *R3-030239,* 3GPP TSG-RAN3 Meeting #34.

[18] Siemens, Measurements for Call Admission Control, R3-030237, 3GPP TSG-RAN3 Meeting #34, http://www.3gpp.org.

[19] 3GPP TS 25.435 v5.5.0, Technical Specification Group Radio Access; UTRAN I_{ub} Interface User Plane Protocols for Common Transport Channel Data Streams, http://www.3gpp.org.

[20] 3GPP TR 25.899 v0.0.3, Technical Specification Group Radio Access Networks; HSDPA Enhancement (Release 6), http://www.3gpp.org.

[21] T. E. Kolding, F. Frederiksen, and P. E. Mogensen, "Performance Aspects of WCDMA Systems with High Speed Downlink Packet Access (HSDPA)," http://www.nokia.com.

Chapter 5

Multiple Antennas

Y. J. Guo and F. C. Zheng

With ever-increasing demand on high-speed data-centric services, it is expected that operators of the third generation mobile communications networks will face capacity problems in a few years. Since spectrum is a limited natural resource, various measures will be taken to improve the spectral efficiency of the air interface, which include spectrum efficient modulation and multiple access schemes, such as the adaptive modulation and coding used for HSDPA and orthogonal frequency division multiplexing (OFDM) to be discussed in Chapter 6. This chapter is focused on the spectrum efficiency enhancement techniques in the space domain. In parallel to frequency, time, and code, space represents another resource and the means of taking advantage of such a resource is to employ multiple antenna systems with antenna arrays either at the receiver or at the transmitter, or both. As with other resources, space can be exploited via space division or space diversity. This divides multiple antenna systems into the following two broad categories.

(1) *Space division-based systems:* They refer to more commonly termed smart antennas which enhance the system performance by directional discrimination, that is, by forming beams towards the desired users and nulls towards the interferers [1–4].
(2) *Space diversity-based systems:* In these systems, the quality of the wireless link is improved by exploiting the space diversity gain. This category includes both the transmit diversity antenna systems and multiple-input multiple-output (MIMO) systems.

Owing to the enormous performance gain promised by multiple antenna systems, the past decade has seen intensive investigation both on smart antennas and on space diversity technologies. As is often commented, space represents probably the last frontier in our endeavor to increase the spectral efficiency of wireless systems. In this chapter, the concepts of smart antennas, transmit diversity antennas, and multiple-input and multiple-output (MIMO) antenna systems are presented. Their advantages and disadvantages are discussed, and practical implementation issues are addressed. Some guidelines for deploying multiple antennas in cellular systems are given.

5.1 SMART ANTENNAS

The capacity of a cellular network depends on many factors, such as the allocated spectrum, the maximum power of the transmitter, and the interference between different users. A common capacity enhancement technique used in the second generation cellular networks is known as sectorization, which is to split an omnidirectional cell into several sectors, typically three. Since the interference between users in different sectors is suppressed by the beam patterns of the antennas installed at the antenna site, the system capacity is significantly increased in proportion to the number of sectors. The basic idea of smart antennas is to replace the antenna for each sector with an antenna array or to replace all the sectorial antennas with a circular array to serve mobile terminals in a 360° angular range. By employing intelligent beamforming algorithms, the antenna beams become adaptive, either providing dynamic sectorization or individual beams for each mobile user.

In an ideal smart antenna system, a beam is formed for each targeted user and the interference caused by other users is significantly reduced by placing them in the direction of nulls or outside of the main beam. It should be self-evident why smart antennas are also referred to as adaptive antennas and beamforming antennas. The main benefits that a smart antenna system can offer include the following:

- Higher system capacity;
- Less power consumption at the mobile terminal and the base stations;
- Higher quality of services such as lower bit error rates and less call drops;
- Greater cell coverage.

As shown in Figure 5.1, the beam formed by a smart antenna system is created by applying different weighting coefficients to different antenna elements that are physically placed apart. The weighting coefficients are produced by virtue of beamforming algorithms according to the signal environment. The weighting operation itself in a smart antenna system can be implemented at radio frequency (RF), intermediate frequency (IF), or baseband. In this chapter, only the RF beaming and baseband (digital) beamforming systems will be discussed due to their wide acceptance by the mobile communications industry.

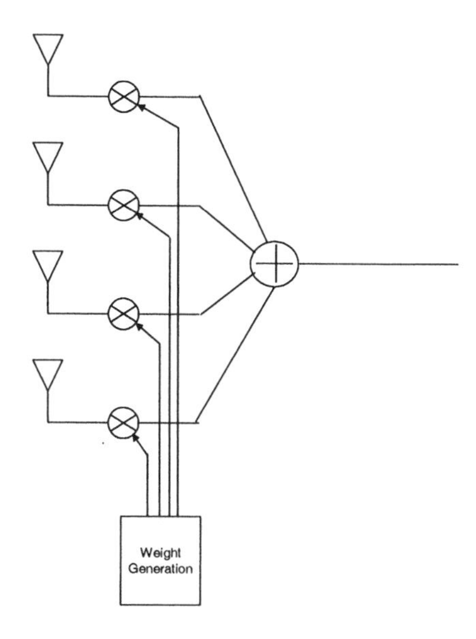

Figure 5.1 Illustration of the smart antenna concept.

5.1.1 RF Beamforming: Adaptive Sectorization

As the earliest beamforming technique, RF beamforming has been around for decades, but it was only recently that we started to see the technique being employed in commercial cellular networks such as the second generation systems to achieve adaptive sectorization [1, 4, 5]. Since the weighting operation is carried out in the RF band and no baseband information is needed, RF beamformer can readily be employed as an add-on appliqué module, which requires no modification to other parts of the existing base stations. This is the reason why RF beamforming tends to be more popular with operators of the second generation mobile communications networks. Nevertheless, RF beamforming can also offer equally significant benefits to the third generation mobile communications networks.

Most sectorized site configurations deployed for the second and third generation mobile communications networks are fixed. However, the fixed uniform sectoring configuration is effective only if the traffic is evenly distributed geographically and stays so all the time. In practice, the traffic load tends to be unevenly distributed and varies with time, which is often caused by the occurrence of hot-spot scenarios such as major sport events and commuter crowds. As an illustration, a more realistic scenario than the uniform user distribution is depicted in Figure 5.2. It is seen in the figure that sector 1 is heavily loaded and probably suffers from call blockage and congestion already, whereas sectors 2 and 3 are only lightly loaded. Apparently, in this scenario, the system capacity cannot be fully utilized and the quality of services would be low.

In fact, the nonuniform traffic distribution shown in Figure 5.2 is very common in real networks [2, 3]. Figure 5.3 shows the traffic load of some real life GSM cells of a major U.K. operator in the Greater London area, where the imbalance in traffic loading among different sectors is clearly demonstrated. It is seen that the traffic load in the busiest cells can be four times as much as that of the quietest cells. A similar pattern also exists for CDMA networks [2]. This traffic imbalance has a major impact on the quality of services of the network since it is the overloaded sectors that experience call drops and blockage, whereas resources allocated to other sectors are simply wasted.

An effective solution to the above problem is to adopt an adaptive non-uniform sectoring strategy, as shown in Figure 5.4. Compared with Figure 5.2, it is seen that the hot-spot traffic is now shared by sector 1 and sector 2 with sector 3 serving a large area of low user density, thus resulting in much more balanced sector loading and much higher system throughput.

In the RF beamforming, sector reconfiguration can be realized physically by adaptively combining or recombining the multiple narrow beams using microwave devices. As a result, almost all parameters related to sectoring can be adjusted that include beamwidth, beam direction, and beam roll-off. The adaptation of these parameters according to dynamic traffic load helps mobile communications network operators to ease the blockage, reduce call drops, and improve the quality of services to the subscribers.

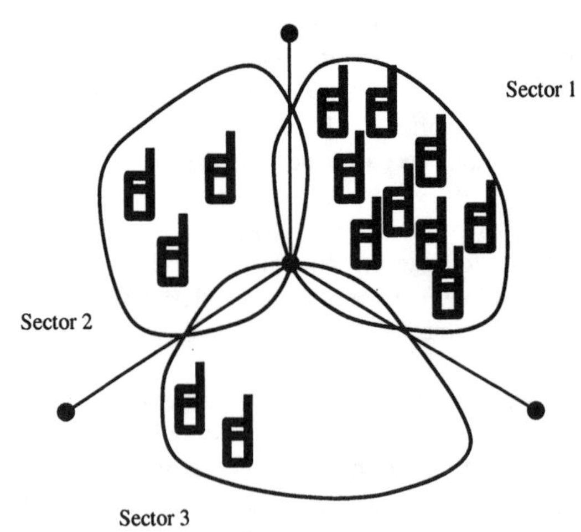

Figure 5.2 Illustration of uniform sectoring for nonuniform traffic.

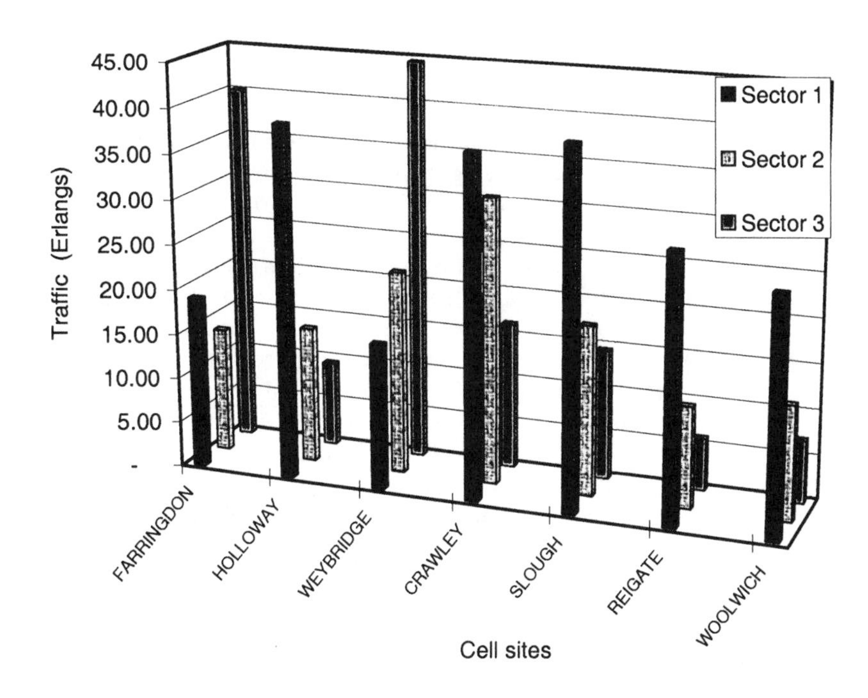

Figure 5.3 The traffic loading of some GSM cells in the Greater London area.

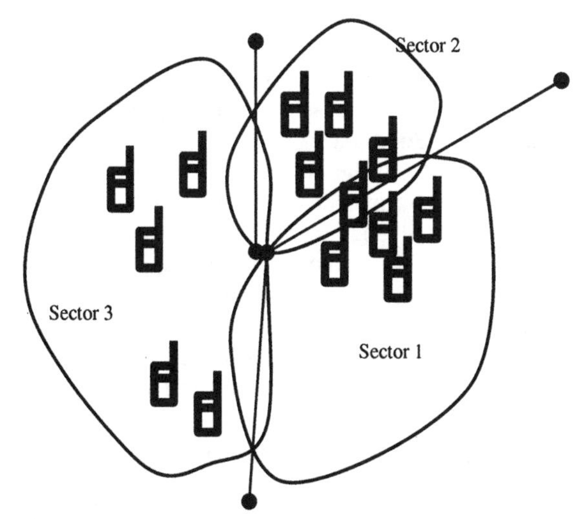

Figure 5.4 Nonuniform sectoring for nonuniform traffic.

The benefits of employing these RF beamforming or adaptive sectorization techniques in cellular mobile communications systems can be summarized as follows:

- *Dynamic sector synthesis and traffic load balancing.* This allows operators to customize sector pointing and beamwidth to balance the total loading across sectors. As a result, the peak loading in the most heavily utilized sectors will be reduced, creating more headroom for traffic growth in the cell and reducing the access failure rate.
- *Handover overhead reduction.* This is due to the fact that RF beamforming normally generates a much sharper sector roll-off than for the conventional antennas. Also, the operators will have the ability to reposition the handover zones from high to low traffic areas, further reducing the handover overhead.
- *Interference control.* Adjustment of antenna gain and phase on a per-beam basis enables operators to tailor the cell's coverage footprint, allowing either for more precise control of pilot pollution from coverage overshoot or for extended coverage required by the local environment.
- *Flexible antenna sharing.* By employing RF beamforming, it becomes possible for different carriers and even different types of systems to share the same antennas while having different sector patterns. For example, varying azimuth and elevation (down-tilt) patterns can be designed for different sections of the band, so operators can share the same antenna and still be able to create or adjust their own sectoring plan.

Since the locations of most hot spots do change with time, sectoring must be changed adaptively through a central controller. However, the rate of change for this purpose tends to be slow. This provides an ideal application scenario for RF beamforming. The RF beamforming can be easily installed as an add-on subsystem. The beamforming circuits for the RF-based smart antennas can be implemented either on the tower top, as with Celleltra's Beamer, or at the tower bottom, as with Metawave's Spotlight.

It should be pointed out that the devices used for RF beamforming, such as attenuators, phase shifters and switches, tend to have limited accuracy and the adaptation of their parameters tends to be slow. Therefore, RF beamforming is mainly useful for realizing low adaptation rate applications such as sector reconfiguration and antenna sharing. Furthermore, at the RF band, there is no direct access to the baseband information such as pilot symbols, so the intelligence level of RF beamforming is rather limited. Thus, RF beamforming per se represents only an intermediate step to smart antennas. By contrast, adaptive digital beamforming, which fully utilizes the baseband information, offers the ultimate solution. Finally, it is worth noting that for an FDMA/TDMA system such as GSM where the frequency reuse factor is less than 1, the use of RF beamforming or adaptive sectorization makes frequency planning more challenging due to the potential cochannel interference that it may cause.

5.1.2 Adaptive Digital Beamforming

By definition, RF beamforming relies on the manipulation of RF signals. Therefore, all user signals using the same RF carrier share the same beam pattern. In digital beamforming, however, beamforming algorithms are applied to the baseband signals directly so an optimum beam can be produced for each individual mobile terminal [1]. To be specific, a narrow beam pointed at each user and nulls in the directions of strong interferers to that user can be formed to maximize the signal-to-interference and noise ratio. As a result, the system can accommodate a much larger number of high and low data rate users. In CDMA systems, every user serves as an interferer to other users in the same cell and adjacent cells, and therefore, adaptive digital beamforming becomes a very effective means of increasing system capacity. For instance, adaptive digital beamformers installed at the base station can create beams with deep nulls in the directions of high data rate users in the uplink. Also, they can produce narrow beams pointed at the high data rate users in the downlink to reduce their interference to other users [6, 7]. Owing to the limitations of RF beamforming, adaptive digital beamforming is now seen as the ultimate solution for smart antennas, although much research is still needed for the downlink.

Before we move on to the next section, it is worth pointing out that smart antennas are mainly a feature for base stations due to the following reasons. First, it is difficult to install an antenna array in mobile terminals because of the limited space. Second, mobile terminals usually are of low height and the angular spread of the received signals can easily be as wide as 360°, which renders smart antennas ineffective. Third, the complexity of the DSP required for beamforming is too high for ordinary terminals considering, among other factors, their limited battery power.

5.1.2.1 Beamforming Algorithms

Given the physical antenna configuration and the radio environment, the performance of an adaptive digital beamforming system is mainly determined by the beamforming algorithm, which generates the weighting coefficients for all the antenna elements. As far as the base station is concerned, the uplink signal is either known or can be estimated. This advantage in the uplink leads to a wealth of beamforming algorithms of different performance and complexity. For the downlink, however, the signal quality at the receiver is not directly known to the base station that acts as the transmitter, so the downlink beamforming is a more challenging task. In this section an overview of the existing beamforming algorithms for both the uplink and the downlink is presented.

Uplink Beamforming

The beamforming algorithms for the uplink fall into three categories: temporal reference (TR) algorithms, spatial reference (SR) algorithms, and blind algorithms. These methods can also be used in a hybrid fashion, thus resulting in the space-time schemes and semi-blind schemes. In the following, the discussions are focused on the TR and SR algorithms due to their relative simplicity.

Temporal reference algorithms The basic concept of this type of algorithm is to choose the antenna weighting coefficients in such a way that the deviation of the array output from a reference signal is minimized. The most commonly used criterion for minimization is the mean square error, which leads to a large group of algorithms termed MMSE (minimum mean square error) [5, 8]. The main advantage of MMSE algorithms is that they can be very simple but effective. The drawback is that some training sequences or pilot symbols may be needed. It is partly for this reason that some special pilot symbols have been designed into the UMTS protocols [9].

The MMSE algorithms can be understood by studying the steady state performance of a beamformer. Assuming there are M antenna elements in a beamformer, the signal input to the beamformer can be represented by

$$X = X_d + X_i + X_n \tag{5.1}$$

where the desired signal vector X_d, the interference signal vector X_i, and the noise vector X_n are given by

$$X_d = [x_{d1}, x_{d2}, \ldots, x_{dM}]^T$$
$$X_i = [x_{i1}, x_{i2}, \ldots, x_{iM}]^T$$
$$X_n = [x_{n1}, x_{n2}, \ldots, x_{nM}]^T$$

respectively. The elements of each vector represent the corresponding values measured at different antenna branches. Assuming the desired signal, the interference signal and the noise are uncorrelated, one has the following correlation matrix [8]:

$$\overline{\overline{R}} = \overline{\overline{R_d}} + \overline{\overline{R_i}} + \overline{\overline{R_n}} \tag{5.2}$$

where

$$\overline{\overline{R_d}} = E\{X_d X_d^H\}$$

$$\overline{\overline{R_i}} = E\{X_i X_i^H\}$$
$$\overline{\overline{R_n}} = E\{X_n X_n^H\}$$

with E denoting averaging operation. Representing the beamformer weight vector by

$$W = [w_1, w_2, \ldots, w_M]^T \tag{5.3}$$

the optimum beamformer weights which results in the minimum mean square error between the reference signal and the array output is given by the so-called Wiener solution [8]:

$$W_{op} = \overline{\overline{R}}^{-1} P \tag{5.4}$$

where

$$P = E\{s^* X\} \tag{5.5}$$

with s representing the reference signal and * denoting the complex conjugate.

In principle, one can choose either a fixed averaging window or a sliding averaging window to calculate the correlation matrix based on the sampled input signals to the antenna branches in order to obtain the optimum beamformer weights by using (5.4). This approach is referred to as "direct matrix inversion" (DMI) and it has the advantage of fast convergence and best performance in a static environment. However, the drawback of the direct matrix inversion method is that its computational complexity is the highest if the number of antenna elements is large. In [10], it was shown how the DMI algorithm could be used to control a two-element array, both in the acquisition stage when the reference signal is available and in the tracking stage when the estimated user signal is used as the reference, and satisfactory results were reported.

In order to reduce the computational complexity of beamforming, a popular approach is to employ the least mean square (LMS) algorithm. The LMS does not require measurements of the correlation matrix, nor does it require the matrix inversion. The standard LMS algorithm is given by

$$W_{n+1} = W_n + \mu \varepsilon_n^* X_n \tag{5.6}$$

where μ is the step-size parameter for adaptation and ε_n is the error signal. In general, the error signal can be expressed as

$$\varepsilon_n = s_n - r_n \tag{5.7}$$

where s_n is the desired signal that can be either a known training sequence for data-aided adaptation or the past decisions for decision-directed adaptation, and r_n is the received signal. The step-size parameter plays an important role in the convergence of the LMS. In general, a large μ should be used in the initial stage of the beamforming to increase the speed of convergence towards the optimum weights and a small μ should be used in the later stage to maintain stability. Since the error signal provides the required feedback signal for adaptation, its content also affects the convergence of the LMS algorithm. LMS possesses the least computational complexity, which is linearly proportional to M, the number of antenna elements, but its rate of convergence could be very slow so some fast weight setup techniques may be needed.

To improve the performance of the LMS algorithm, the normalized least mean square (NLMS) algorithm can be used, which offers more stability and greater convergence speed. The NLMS algorithm is given by

$$W_{n+1} = W_n + \mu\varepsilon_n^* \frac{X_n}{\|X_n\|^2} \tag{5.8}$$

From (5.8), it can be seen that the computational complexity of the NLMS is only slightly higher than that of the LMS due to normalization. A method to further speed up the convergence is to employ the recursive least mean square (RLS) algorithm given by

$$W_{n+1} = W_n + \zeta_{n+1}^* K_{n+1} \tag{5.9}$$

$$\zeta_{n+1} = s_{n+1} - W_n^H X_{n+1} \tag{5.10}$$

where H represents transpose conjugate. Although (5.9) and (5.10) are similar to (5.6) and (5.7), the calculation of the gain factor K_{n+1} in (5.9) is relatively complex, which results in much increased computational complexity of the RLS compared with the LMS. The basic form of the RLS algorithm suffers from numerical instability: The low numerical accuracy incurred in every iteration may eventually change the non-negative nature of the covariance matrix, thus resulting in unreliable results. Among a number of modified versions of the RLS, the QR-RLS is known to have superior numerical stability and lower complexity [8]. As shown later, however, its computational complexity is higher than RLS when M is small.

Spatial reference algorithms This type of algorithm is aimed at estimating the angle of arrival (AoA) of the incoming signal first. Then, a beam is formed in the desired direction, typically the same as AoA [11]. Naturally, the focus of the signal processing for spatial reference algorithms is on the AoA estimation and there are several methods available in open literature that include MUSIC and ESPRIT [1]. In the following, a less well-documented but simpler method based on the fast Fourier transform (FFT) is described.

The output of an equally spaced and uniformly excited linear array antenna is given by

$$y = \sum_{n=0}^{M-1} x_n e^{j2\pi nd \sin\theta_i / \lambda} \tag{5.11}$$

where x_n is the signal at the nth element, d is the interelement spacing, θ_i is the "look-direction" of the antenna, and λ is the operating wavelength. If the following holds

$$d \sin\theta_i / \lambda = i / M \qquad (i = 0,1,\ldots,M-1) \tag{5.12}$$

or

$$\theta_i = \sin^{-1}\left(\frac{i\lambda}{Md}\right) , \tag{5.13}$$

(5.13) can be rewritten as

$$y_i = \sum_{n=0}^{M-1} x_n e^{j2\pi ni / M} \tag{5.14}$$

(5.14) is of the form of discrete Fourier transform (DFT) which can be computed by the fast Fourier transformation (FFT) technique. The implication of the above mathematics is that, given the interelement spacing d/λ and the number of antenna elements M, performing the DFT on the outcome signal of the array will give the output of M spatial beams pointed at θ_i ($i = 0, 1, \ldots, M-1$). For instance, if $d/\lambda = 0.5$ and $M = 4$, (5.14) can be written as

$$y_i = \sum_{n=0}^{3} x_n e^{j\pi n(i-1)/2} \tag{5.15}$$

and y_i gives the output of four fixed beams pointed at $\theta_0 = -30°, \theta_1 = 0$, $\theta_2 = 30°$, $\theta_4 = 90°$.

The approach of AoA estimation using FFT can be summarized as follows:

1. Sampling the input signal at all elements;
2. Using FFT to obtain the spatial spectrum of the signal;
3. Finding the spectrum peak with index i_0;
4. Calculating the weight of the antenna by using

$$w_{i0} = e^{-j2\pi m i_0 / M} \tag{5.16}$$

It may happen that there is no significant peak in the spatial spectrum. In that case, an interpolation method can be used.

Table 5.1

Complexity Comparison of Different Beamforming Algorithms

Algorithm	*Products*	*Sums*
LMS	8M + 2	8M
NLMS	10M + 7	10M + 6
RLS	$8M^2 + 8M + 7$	$6M^2 + 7M + 6$
QR-RLS	$5M^2 + 42M + 23$	$2M^2 + 32M + 21$
FFT	$4M\log_2 M$	$2M\log_2 M$
IBS	4M + 1	2M
DMI	$2M^3/3 + 2M^2$	$M^3/3 + 2M^2$

Computational complexity One factor to consider in choosing the right beamforming algorithm is the computational complexity. To this end, Table 5.1 gives a complexity comparison of different beamforming algorithms. For the direct matrix inversion method, it is assumed that Cholesky decomposition technique is used. The IBS algorithm will be discussed later in this section. It is seen from Table 5.1 that the computational complexity of each algorithm depends strongly on the number of antenna elements M. For instance, although the number of operations required by DMI is on the order of M^3, it is actually smaller than QR-RLS when M is small. Also, the QR-RLS is simpler than RLS only for large M. Overall, LMS and IBS are the simplest.

Downlink Beamforming

Adaptive digital beamforming can provide significant benefits in both the uplink and the downlink. Since the uplink beamforming weights can be derived directly from the received signals, the beamforming process for the uplink is far more straightforward than that for the downlink. For time division multiplexing (TDD) systems, there is a high degree of uplink-downlink reciprocity, so the weighting coefficients obtained for the uplink can possibly be used for the downlink. For frequency division multiplexing (FDD) systems, however, the fading profiles and array responses are largely uncorrelated between the uplink and the downlink [11], so the beamforming weights obtained in the uplink cannot be reused in the downlink directly.

Downlink beamforming is complicated further by the fact that it is inherently different from uplink beamforming. For the uplink, an optimum beam can be formed for each individual mobile terminal to maximize the signal-to-interference and noise ratio (SINR) observed at the output of the beamformer. For the downlink, however, the quality of the signal received by each mobile terminal depends not only on the beam pattern formed by the smart antenna at the base station for the interested mobile terminal and the associated transmit power, but also on the beam patterns formed for other mobiles and the associated powers. In theory, therefore, all the beams should be jointly optimized in association with the transmit powers in order to achieve maximum SINR at each mobile and minimum interference to other users. Unfortunately, such a global optimization approach may not be feasible in practice due to both the excessive signal processing complexity and the inaccuracy of downlink AoA estimation. Therefore, the following practical approach could be used for downlink beamforming. First, based on the signal received from the uplink, the angles of arrival (AoAs) of all the signal components for each user are estimated. Second, based on the AoAs of each intended user, a common beam is synthesized for it. It should be pointed out that although the propagation mechanism for the uplink is similar to that for the downlink, the actual dominant paths are generally different in the two cases because of the frequency difference in the FDD scheme. Therefore, the common beam for each user should cover all the significant multipath signals in different directions.

Iterative Beam Steering[1]

As discussed above, temporal reference algorithms are aimed at finding the optimum beamforming weights without specific regard to the AoA of the signals. In contrast, spatial reference algorithms are aimed at estimating the AoA of

---

[1] Dr. Guo developed the IBS algorithm when he was with Fujitsu Europe Telecom R&D Center in England.

incoming signals and then synthesizing a beam accordingly. Among the temporal algorithms, iterative algorithms such as LMS and NLMS are most popular due to their low computational complexity, but they do not provide the information on the AoA so some parallel algorithms may be needed for downlink beamforming. In the following, an iterative beam steering (IBS) algorithm is presented. The IBS algorithm has the advantage of lower computational complexity than the LMS and producing the AoA information directly in the meantime.

Beam steering is similar to the phased array approach used in the military. Its main objective is to create a beam of almost constant shape to follow the intended signal. As such, it is inherently a suboptimum solution in theory, as no nulling is intended in the direction of interferers. In practice, when only a small number of antenna elements are used in a well-loaded system, nulling against a large number of interfering signals, which can be achieved by using the optimal beamforming technique may become ineffective. Therefore, the performance of the beam-steering technique could be very similar to that of the optimal beamforming. Since the beam-steering technique can normally provide some information on the angles of arrival (AoA) of the wanted signal, which can subsequently be used in the downlink, it stands as a strong candidate for the application in the mobile cellular communications systems.

The essence of IBS algorithm is to produce a steering beam in an iterative manner. Mathematically, the basic IBS algorithm is given as follows:

$$\mathbf{w}_n = \alpha \mathbf{w}_{n-1} + (1-\alpha) s_n^* \mathbf{w}_n \tag{5.17}$$

where $\mathbf{w}_{n-1}$ and $\mathbf{w}_n$ are the old and new beamformer weights, respectively, s_n is the reference signal, and α is the forgetting factor that is used to control the effective window size of the averaging process. Compared with (5.6) for LMS algorithm, a major difference in (5.17) is that no error signal is needed for the IBS. In Appendix 5A, it is shown that the IBS algorithm produces an estimate on the weighting coefficients of an ideal steering beam in an AWGN channel.

For practical implementation, two different versions of the IBS algorithms can be employed. The first one is a straightforward implementation of (5.17) as

$$\mathbf{w}_n = \alpha \mathbf{w}_{n-1} + \frac{(1-\alpha) s^*{}_n \mathbf{x}_n}{\|\mathbf{x}_n\|^2} \qquad for \quad 0 < \alpha < 1 \tag{5.18}$$

where the normalization of the input signal in the second term is introduced to make the algorithm insensitive to other operations in the receiver. An advantage of using (5.18) is that since the information on the channel phase is not needed, the delay introduced by the system is minimized. To obtain a better steady-state performance in a fading channel, however, the following version can be used to make use of the channel phase information ϕ obtained by the channel estimator:

$$\mathbf{w}_n = \alpha \mathbf{w}_{n-1} + \frac{(1-\alpha)e^{-j\phi} s^*{}_n \mathbf{x}_n}{\|\mathbf{x}_n\|^2} \qquad \textit{for} \quad 0 < \alpha < 1 \tag{5.19}$$

If the phase shifts introduced by fading are uncorrelated with those introduced by the AoA change of the wanted user, the estimation of the fading and AoA components are decoupled, so the IBS is required to track only the AoA of the signal component of interest, thus resulting in better performance.

The iterative beam steering algorithm is based on the correlation between the signal at the input of the antenna and a reference signal. When applied to UTRAN, the reference signal can be taken as the control signal used in the dedicated physical control channel (DPCCH) [9].

The advantages of the IBS can be summarized as follows. When the IBS is employed, the AoA information can be extracted from the reverse link antenna weights and applied to the downlink beamformer. The IBS operates in an iterative fashion and it does not require an error signal, so the system delay and the buffer size required in the hardware can be reduced. Compared with other existing optimal beamforming and beam-steering algorithms, the IBS has the advantages of fast convergence, high stability, and low complexity. It is even marginally simpler than the celebrated normalized least mean square (NLMS) algorithm, but can converge within a few symbols even in heavily loaded systems.

5.1.2.2 Smart Antennas for CDMA

In a CDMA system, there are a large number of users sharing the same carrier frequency and time slot. Therefore, employing smart antennas can increase system capacity by suppressing interference caused by other users. If all the users in a CDMA system are of the same data rate as in IS-95 systems, the benefits of employing four antennas in each sector of a three-sector site could be comparable to employing six-sector sectorization, and one may argue that it is necessary to employ much larger arrays. However, both UTRAN and CDMA2000 1x are multimedia and thus multirate systems. The user data rate can vary from tens of kilobits per second to several megabits per second. In such a system, the nulling effect offered by adaptive antennas can be significant. With an array of M elements, there are $M-1$ degrees of freedom and thus $M-1$ independent nulls can be formed. Since each high data rate user is equivalent to many low data rate users, nulling a few of them would lead to great increase in system capacity. To incorporate adaptive digital beamforming, both WCDMA and CDMA2000 have provided some special signaling hooks. In particular, adaptive beamforming has been studied by 3GPP as an optional feature.

A powerful device in CDMA systems is the so-called Rake receiver, which is used to collect energy carried by different multipath signals by despreading different spread spectrum signal components and combining them in a constructive manner [12]. For the uplink, if the multipath signals have a small angular spread compared with the likely beamwidth, a single beamformer can be used before the combiner. Then the beamformer will create a common beam for all the dominant multipath signals. This type of beamformer is called common beamformer. By the likely beamwidth, we mean the beamwidth associated with the array with uniform weights. On the other hand, if the multipath signals have a large angular spread compared with the likely beamwidth, one can employ the so-called finger beamformer, which is to employ a beamformer for each finger of the Rake receiver and place the combiner behind them. The finger beamformer is sometimes called the two-dimensional Rake receiver. A generic architecture of the uplink finger beamformer using the normalized least mean square (NLMS) algorithm [8] is shown in Figure 5.5, where each finger beamformer is equipped with a beamforming engine based on the NLMS algorithm and a channel estimator for phase estimation and correction. In Figure 5.5, the matched filter (MF) blocks refer to correlators [7]. A combination of the pilot signal and tentative decisions is taken as the common reference signal.

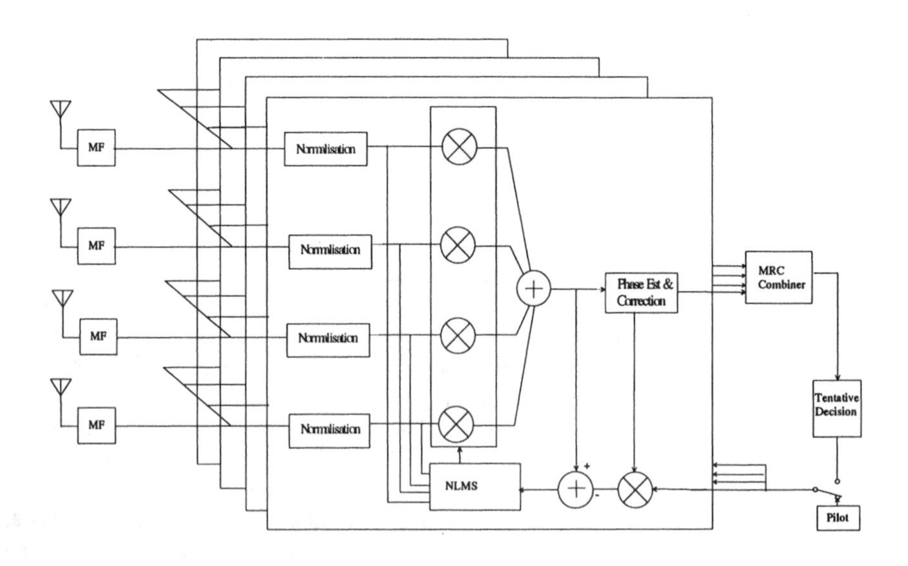

Figure 5.5 NLMS-based uplink finger beamformer for UTRAN.

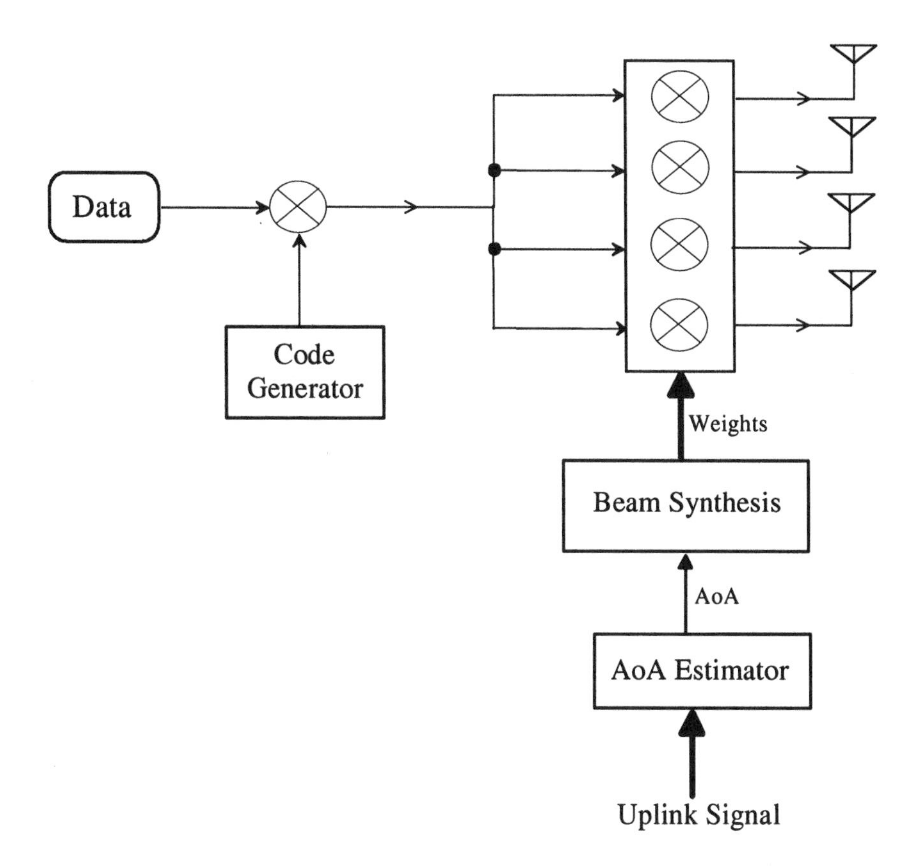

Figure 5.6 Illustration of the downlink beamformer.

It is known that the NLMS algorithm may be too slow to track the rapid fading of radio channels in mobile environments. To overcome this difficulty, two countermeasures are introduced in Figure 5.5. First, signals at the output of the correlators are normalized to the mean signal intensity of each individual path before being fed into the beamformer. Second, the reference signal is phase-compensated before being subtracted from the output of the beamformer. These two functions serve to separate the fading compensation from the angular tracking of mobiles, thus improving the performance of the NLMS algorithm.

Based on the concept presented in Section 5.1.2.1, a generic diagram of the downlink beamformer is shown in Figure 5.6. Compared with the uplink beamformer shown in Figure 5.5, the configuration of the downlink beamformer appears to be much simpler, which is due to the fact that the major signal processing task is performed by the AoA estimator. As discussed earlier, when the

LMS type of algorithms is used, another algorithm for AoA estimation may be needed for downlink beamforming. Alternatively, one may employ the IBS algorithm for uplink beamforming as shown in Figure 5.7. In that case, no other AoA estimation algorithm is needed for the downlink beamforming, as the AoA information can be extracted from the beamforming weights.

5.1.2.3 Simulation Results for WCDMA

In order to obtain a quantitative capacity measure of the UTRAN system employing adaptive antennas, a four-element linear adaptive antenna array is studied for UTRAN base stations with three sectors in macro-cellular environments, using the parameters specified in the 3GPP documents [7]. In the simulations, it was assumed that all the users are randomly located according to a uniform distribution in each sector, and the multipath components of each user signal are spatially distributed according to a Gaussian distribution with 2.5° standard deviation. The ITU Vehicular B model is used for all the radio channels and the Doppler frequency is assumed to be 80 Hz. The processing gain is chosen as 128, which corresponds to a 32 kbps raw data rate. A 1/3 rate convolutional coding scheme with constraint length $K = 9$ is used and the decoder employs soft decision Viterbi algorithm [12].

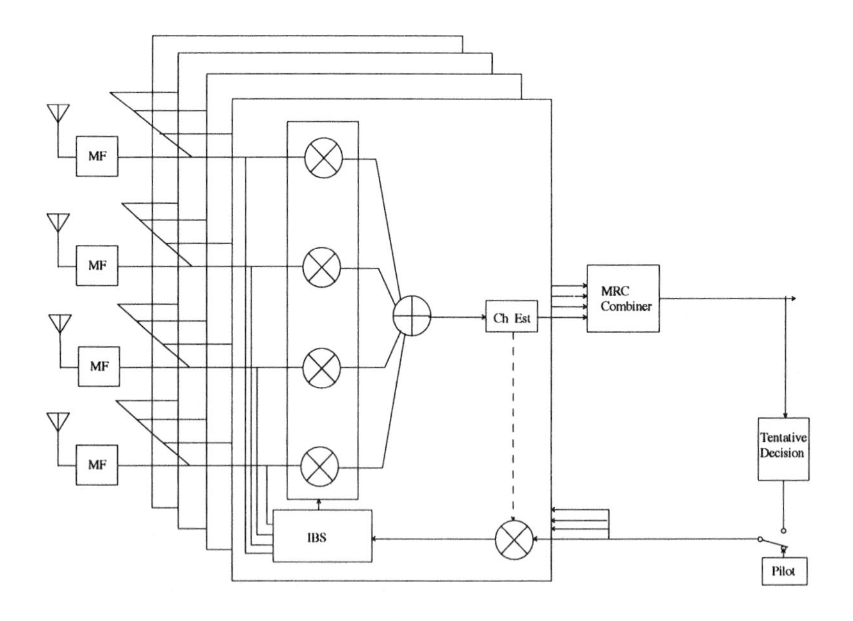

Figure 5.7 An illustration of IBS-based uplink beamformer.

Figure 5.8 shows the required signal-to-noise ratio per information bit (E_b/N_o) per antenna branch in the uplink for different numbers of simultaneous speech channels which the adaptive antenna array can support in an isolated sector to achieve the targeted bit error ratio (BER) of 10^{-3}, where a practical antenna pattern with –3-dB tapering at the sector boundaries is used. The benefit of using thermal noise as the reference is that E_b/N_0 per antenna branch is directly related to the transmit power of the mobile. To account for discontinuous transmission of voice signals, a voice activity factor (VAF) of 0.5 has been assumed. For comparison, simulation results obtained using one antenna and two branch diversity are also shown in the figure. It is seen that when only one antenna is used, the maximum number of speech channels which the system can support is about 80. The system capacity is increased to about 160 when two-branch diversity is employed. Using the adaptive antenna array, the maximum number of simultaneous speech channels the system can support reaches about 320. When the intercell interference is considered, it is expected that the average system capacity will become much smaller than that shown in Figure 5.8 with about 35% reduction, but the capacity achieved by using the adaptive antenna is still approximately four times that achieved using only one antenna. Within a certain limit, this capacity gain increases proportionally with the number of antenna elements used.

Figure 5.8 also reveals the advantage of using adaptive antennas in saving mobile transmit power. It is observed that in a system loaded with 160 speech users, using a two branch spatial diversity antenna at the base station requires more than 14 dB E_b/N_0 to achieve a BER of 10^{-3}, whereas with the adaptive antenna array of four elements only 1 dB E_b/N_0 per branch is required to achieve the same BER. This implies that the transmit power of the mobile terminals can be reduced by about 13 dB, thus resulting in a many fold increase in battery life. Alternatively, the communication range of the uplink could be doubled under the same load.

The mobile transmit power saving achieved in different loading conditions when using the adaptive antenna at the base station is shown in Figure 5.9. It may be surprising to note that the gain obtained by using the four-antenna array can be much larger than the familiar figures, that is, 6 dB against using one antenna element and 3 dB against using two-branch diversity antennas. This is due to the fact that once a system with two-branch diversity or one antenna is fully loaded, there are few additional users that can be admitted by increasing the transmit power of the mobiles, but using the four-element adaptive antenna array can push up the saturation point substantially.

As discussed earlier, a major advantage of using adaptive antennas is to accommodate high data rate users. The adaptive antenna not only can support a much greater number of simultaneous high data rate users by virtue of a narrow beam, but can also reduce the strong interference caused by high data rate users by nulling, thus increasing the overall system capacity. To illustrate this point, the following scenario was studied.

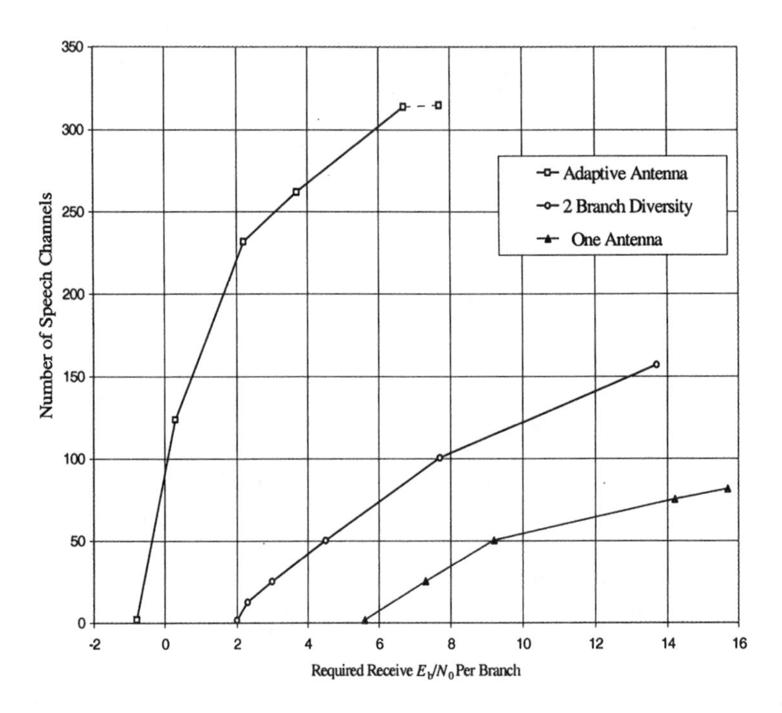

Figure 5.8 Capacity comparison of using one-antenna, two-branch diversity and the adaptive antenna in the uplink for the chosen scenario.

Assume that there is a 0.5 Mbps high data rate user located at –10° and a speech user located at 10° with 32 Kbps data rate. To the speech user, the high data rate user is equivalent to a group of 30 colocated speech users with a voice activity factor of 0.5 interfering with it. When the adaptive antenna array is used for the speech user, an optimum beam pattern can be so formed that the interference from the high data rate user can be reduced to a negligible level. The beam pattern formed for the speech user is shown in Figure 5.10 and it is observed that a deep null is formed in the direction of the high data rate user and the main beam is also shaped. Figure 5.11 shows the BER for the speech user achieved by using the adaptive antenna; also included is the BER achievable for a single user when there is no multiuser interference. It can be seen that, thanks to the adaptive antenna array, the E_b/N_0 required for the speech user to achieve the targeted BER of 10^{-3} in the presence of a high data rate user is more or less the same as that for the single user case. In practice, this means that more active users can be accommodated in the sector, or the power required for the speech user can be reduced. Undoubtedly, the null depth that can be realized in practice is less than that shown in Figure 5.10 due to the performance limitation and the tolerance of various analog devices, including antennas, but even a null with moderate depth, say, –15 dB, means that the adaptive antenna can easily accommodate an order of magnitude increase in the data rate of some services.

5.1.3 Antenna Configuration

Another choice in the design of smart antennas is the configuration of the antenna array. Most conventional cellular systems are based on the sectorized configuration, in which each omnidirectional cell is divided into three or six angular cells and these cells are served by the base station equipment located at the bottom of the antenna tower. A natural evolution from such a configuration into smart antennas might be to introduce a linear array into each sector and the advantage is that the handover technique implemented in the sectorized configuration can be expanded without radical changes. The disadvantage is that the total number of antennas used at each site will be significantly greater.

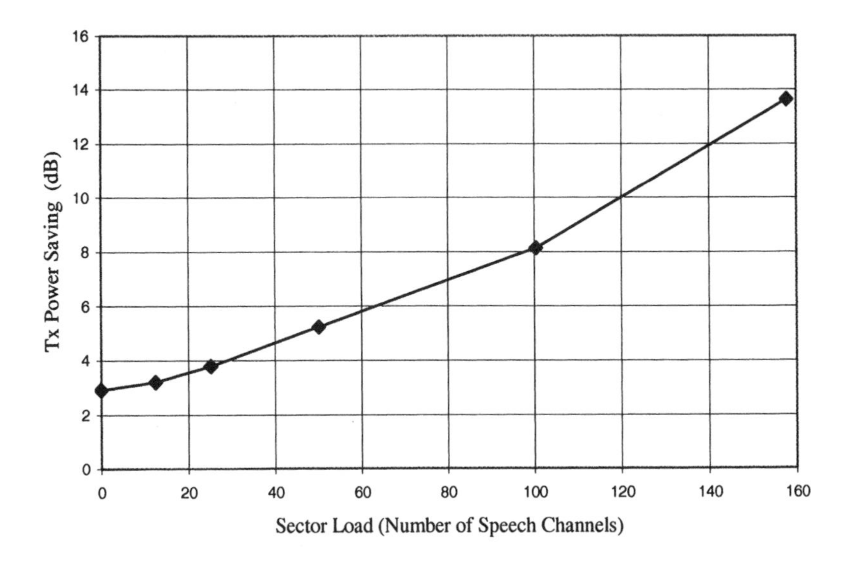

Figure 5.9 Mobile transmit power saving in different loading conditions when using the adaptive antenna array at the base station compared with using two-branch spatial diversity.

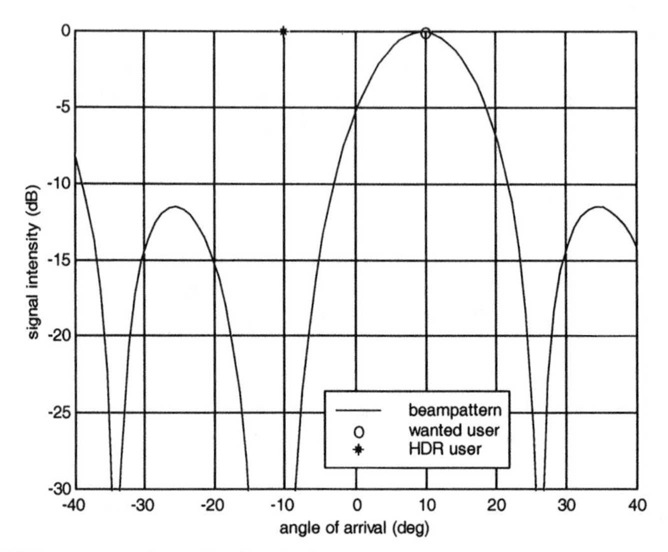

Figure 5.10 Beam pattern formed by the adaptive antenna array.

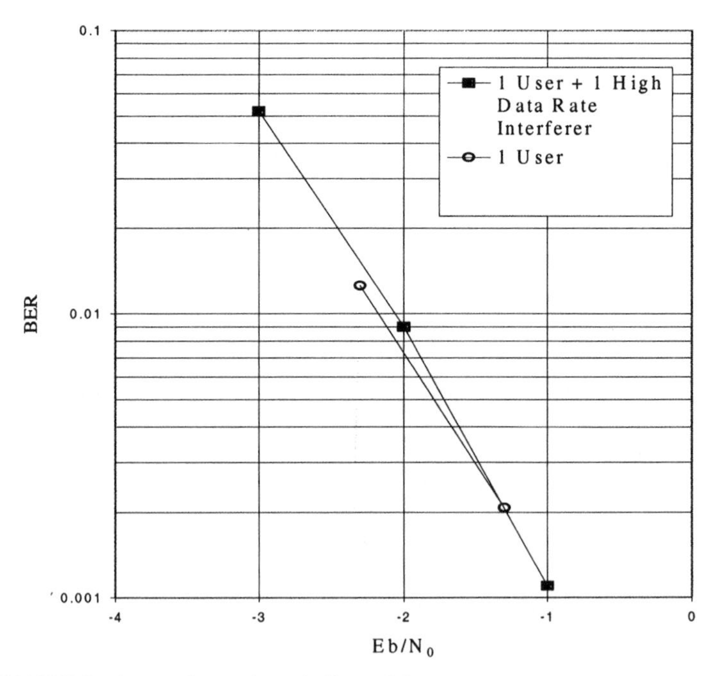

Figure 5.11 BER for the speech user shown in Figure 5.6.

The performance of an adaptive antenna system depends on the number of antenna elements used in the array. In theory, a larger number of antenna elements are generally preferable to achieve maximum benefit. Because of such factors as space, economy, and limitation of the processor speed, however, it is expected that most linear adaptive antenna arrays employed in future mobile communications systems will consist of between four to eight elements in a sector.

The beamwidth of a linear antenna array is inversely proportional to the length of the array. When the number of antenna elements in an array is small, the length of the array can be increased by increasing the interelement spacing in order to produce a narrow beam. On the other hand, increasing the interelement spacing will produce so-called grating lobes if the elements are equally spaced, which means that when creating a beam (main lobe) in one direction, a similar beam automatically appears in another undesired direction. For an array with a few elements, this problem can be eased to some limited extent by using unequally spaced arrays, but, generally speaking, a trade-off between the beamwidth and the level of grating lobes is needed to optimize the system capacity. Another factor which deserves attention when optimizing the antenna spacing is the coupling between antennas. Although the antenna coupling can be incorporated in the beamforming algorithm, it may prove preferable in practice to keep the coupling to a negligible level, which determines the minimum interelement spacing. As a rule of thumb, the interelement spacing should be not too greater than a half-wavelength.

Figure 5.12 shows a linear antenna array used for one of Fujitsu's smart antenna testbeds [7]. It is seen that the physical antenna consists of four panels. Each panel holds two subarrays of four horizontally polarized and four vertically polarized printed elements, respectively. The antenna pattern of each subarray is fixed and beamforming is only performed by treating each subarray as an antenna element.

An alternative approach to using a linear array is to employ elements radiating in all 360° directions in the horizontal plane and arrange them, say, in a circle, which results in a circular array. Since each antenna serves all the mobile terminals in all directions, the number of antennas serving each user and therefore the antenna gain potentially available are much greater than that using several linear arrays. An important advantage of this configuration is that a much smaller number of antenna elements would be needed to achieve a given capacity. This has a profound effect on the cost of the smart antenna system. The power amplifiers are significant cost contributors to downlink smart antennas and each antenna branch requires a power amplifier. Therefore, using a single array means that smart antennas become much more affordable.

5.1.4 Practical Issues

In order to apply the smart antennas technology to commercial mobile communications networks, some practical issues that include calibration, cost control, and deployment strategy must be resolved first.

5.1.4.1 Calibration

Owing to mechanical and electrical variations of analog components, such as amplifiers and cables, signal branches of an antenna array tend to have different transfer functions. Therefore, system calibration is needed to compensate for the phase errors $\Delta\phi$ and magnitude errors Δa at each branch. In general, two types of calibrations are needed: off-line calibration and online calibration [13-15]. The offline calibration is done prior to system installation to combat component variations produced in the fabrication process, whereas the online calibration is performed when the system is in operation in order to correct magnitude and phase errors occurring from temperature variations, aging, and component replacement. The post calibration targets for these errors were specified as $\Delta\phi<3°$ and $\Delta a<0.5$ dB in the TSUNAMI project [15]. Obviously, the online calibration must be carried out periodically, so a calibration circuit must be built into such an adaptive antenna system.

Figure 5.12 An antenna array used in an adaptive antenna system.

The basic principle of calibration is to inject a calibration signal into every antenna branch, or signal path. Since all the calibration signals injected into all the branches are the same, the phase, magnitude, and delay errors can be obtained by comparing the signal outputs at the other ends and compensations can be introduced accordingly.

The calibrations for transmission and reception are normally carried out in different ways. For reception, the same calibration signal can be fed into the antenna branches by using power splitters and directional couplers, which is done when the adaptive antenna is idle. For transmission, the live data can be used as the calibration signal, but in this case the RF signal should be downconverted back to the baseband.

For uplink beamformers based on temporal reference algorithms discussed in Section 5.1.2.1, calibration is not required as the beamforming algorithms can compensate for the nonuniformity of the signal paths. For adaptive antennas based on beam steering and the estimation of the angle of arrival of multiuser signals, however, it is required that all the antenna branches have the same transfer function so calibration is needed. Generally speaking, calibration is needed for downlink beamforming.

5.1.4.2 Cost and Deployment Strategy

The RF cost has served as the major factors affecting the wide acceptance of smart antennas. In a conventional base station configuration, the power amplifier, which accounts for between 30% to 40% of the total base station cost, is located in a rack for base station equipment. For downlink, signals from the power amplifier are carried over RF cables to the antenna mounted at the top of the antenna tower. For the uplink, the signals received by the antenna are carried down to the RF devices via RF cables. Admittedly, it is relatively easier to mount the low-noise amplifier and RF filters next to the antenna due to the form factor and power consumption.

The RF cables between the antenna and power amplifiers introduce some losses, which in turn make the power amplifier more expensive as greater power is needed. The RF cables are expensive and heavy and need strong mechanical support. Also, the RF cable is regarded as one of the common causes for mechanical failures in base stations. To employ smart antennas, it appears that one needs to increase the number of power amplifiers and the number of cables, which may make the base station much more expensive and more vulnerable to mechanical failures. This calls for the application of radio over fiber technology. By placing the power amplifier next to the antennas and replacing the RF cables with optical fibers, the RF losses in the cables are avoided, so lower power output is required of the power amplifiers, thus reducing the cost and size of the power amplifiers. Also, optical fibers are more reliable and cheap. All these factors make smart antennas much more economical than usually perceived.

Adaptive antennas can be used to increase either cell coverage or system capacity. For the latter application, the cell range is increased by virtue of the array gain instead of the transmit power increase at the base station. For the third generation mobile communications networks, a stepwise strategy for deploying smart antennas has been proposed by some network vendors. The first step is coverage-driven and is used for the initial network deployment, in which the smart antennas are used only in the uplink to reduce the number of base station sites. This is because in a lightly loaded network, the cell range is mainly limited in the uplink by the power of mobile terminals. In this scenario, the cost of smart antennas can be easily offset by saving in the base station equipment and site leasing cost. In the second step, when the traffic load in the network becomes heavy and the downlink power becomes the capacity bottleneck, smart antennas in the downlink can be introduced.

It should be pointed out that smart antennas are regarded as too expensive mainly because of the cost of power amplifiers. In fact, owing to the antenna gain provided by the antenna array, power amplifiers with much smaller power output will be needed. This may result in comparable or even lower costs comparing with sectorized configuration.

5.2 TRANSMIT DIVERSITY ANTENNAS

Mobile radio channels suffer from a phenomenon known as fading. Every signal received at the receiver comprises multiple components of the same signal caused by reflection, refraction, and scattering, normally within a certain angular range. Owing to the movement of the mobile users and the change of the surrounding environment, these components vary from in phase to out of phase, and therefore the strength of the resultant signal at any given point, moving or stationary, fluctuates between highs and lows. The fading profiles of such signals depend on many factors, but they tend to become uncorrelated if the points of measurements are distant. The consequence is that when transmit or receive antennas are placed significantly far apart, signals transmitted or received become uncorrelated or weakly correlated. Naturally, if multiple antennas are placed at locations where the fades of the signal are independent, then the combined signal quality can be well maintained, as it is very unlikely for all the signal paths to fade simultaneously. Receive diversity antennas (or diversity-combining antennas) have been known for a long time and they are aimed at exploiting the feature of uncorrelated radio channels by combining signals at the output of different antennas [16]. The most popular receiver diversity technique is called maximum ratio combining (MRC), which weights signals from different antenna branches according to individual signal to noise ratios in order to achieve the maximum signal-to-noise ratio (SNR) for the combined signal [16].

Although it is a common practice to employ two-branch receive diversity at the base stations, it is difficult to accommodate more antenna branches at the mobile terminals due to restrictions on space and power consumption. For this reason, the transmit diversity (TD) scheme using multiple antennas has become a popular research topic in recent years. Its principle is to use multiple transmit antennas in such a way that the signals transmitted by different antennas have independent fading profiles at the receiver, so diversity can be achieved by using only one antenna at the mobile terminal. Transmit diversity antennas are particularly effective in CDMA systems where the Rake receiver can distinguish different multipath components and add them in a constructive manner. The concept of transmit diversity antennas was proposed in [17] for two antennas and generalized in [18] to an arbitrary number of antennas.

To achieve a meaningful diversity, the signals transmitted from different antenna elements should go through independent fading. As such, the interelement spacing in a transmit diversity antenna should be, in contrast to the beamforming system, as large as possible. In a typical macro-cell scenario, at least eight wavelengths are needed.

5.2.1 Space Time Block Coding

A key component in a transmit diversity system is the space-time encoder. The simplest form of the space-time encoder can just be a data splitter, in which case the same data sequence is transmitted by all the antenna elements. At the receiver, the maximum ratio combining or other combining techniques can be used to recover the transmitted sequence. A more elegant approach is to apply space and time block coding (STBC) to the transmitted signal. The conventional STBC scheme for two transmit antennas (2-Tx) was originally proposed by Alamouti in [17]. As is depicted in Figure 5.13, the complex symbol sequence is first parsed into code vectors $S = [s(2n), s(2n+1)]$. S is then encoded in the following fashion before going to the next stage of the transmission chain: the first antenna branch transmits $s(2n)$ and $-s^*(2n+1)$ over two symbol periods and the second antenna branch transmits $s(2n+1)$ and $s^*(2n)$ over the same two-symbol periods, where the superscript * denotes conjugation.

Letting the received signal at time instant l be $r(l)$, the corresponding additive Gaussian noise be $v(l)$, and the channel gain from antenna i to the receiver be $h_i(l)$, one has

$$R = \overline{\overline{H}}S + V \tag{5.20}$$

where

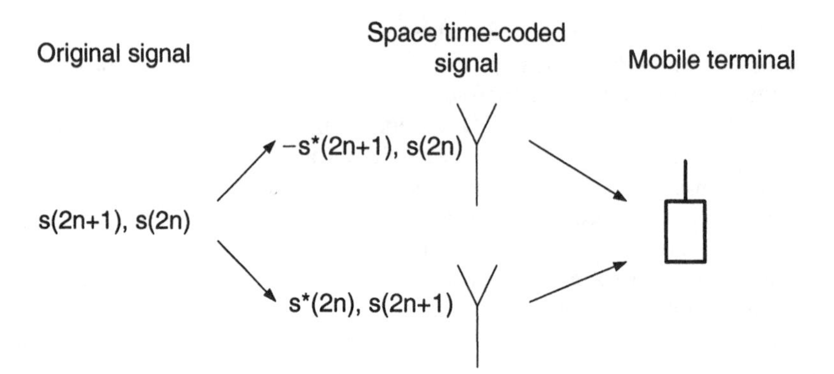

Figure 5.13 Illustration of a two-branch transmit diversity antenna using STBC.

$$R = [r(2n), r^*(2n+1)]^T$$
$$V = [v(2n), v^*(2n+1)]^T$$
$$\overline{\overline{H}} = \begin{bmatrix} h_1(2n) & h_2(2n) \\ h_2^*(2n+1) & -h_1^*(2n+1) \end{bmatrix}$$

If the channel is quasi-static [17], one has

$$h_i(2n) = h_i(2n+1) = h_i \qquad i = 1,\ 2$$

which leads to

$$\overline{\overline{H}} = \begin{bmatrix} h_1 & h_2 \\ h_2^* & -h_1^* \end{bmatrix}. \tag{5.21}$$

The channel matrix given in (5.21) is unitary, which means that it has the special property that its inverse is the same as its conjugate transpose. Applying the conjugate transpose of the channel matrix in (5.21) to (5.20) leads to

$$Q = \begin{bmatrix} |h_1|^2 + |h_2|^2 & 0 \\ 0 & |h_1|^2 + |h_2|^2 \end{bmatrix} S + \overline{\overline{H}}^{+} V \tag{5.22}$$

where

$$Q = [q(2n), q(2n+1)]^T \quad (5.23)$$

From (5.22), it is seen that now the space and time coded signal $s(2n)$ and $s(2n+1)$ transmitted by the two antennas are separated in the received signal $q(2n)$ and $q(2n+1)$. In other words, $q(2n)$ is dependent only on $s(2n)$ and $q(2n+1)$ is dependent only on $s(2n+1)$. However, the quality of the two received signals is dependent on the quality of the two independent channels. This means that employing the STBC at the transmitted antenna leads to diversity gain at the receiver. This simplification in the space time decoder is due to the coding structure shown in Figure 5.13. It can be proved that the above STBC scheme yields the same diversity gain as the maximum ratio combining (MRC) at the receiver, but the latter requires two receive antennas at the mobile terminal, thus incurring a higher handset cost. This is, in fact, one of the major motivations for the original Alamouti STBC scheme proposed in [17].

The above Alamouti scheme for two transmit diversity antennas can be extended into four transmit diversity antennas to increase the diversity order, that is, the number of independent channels. In that case, however, there will be a rate loss of 50% because only four symbols can be transmitted in an eight-symbol period [18].

5.2.2 Space Time Transmit Diversity

The two-branch transmit diversity antenna technique has now been incorporated into the specifications of the UTRAN air interface under the name of space time transmit diversity (STTD). The space time transmit diversity scheme is slightly different from the two branch space time block coding (STBC) scheme. Assuming that the data bits to be transmitted are b_0, b_1, b_2, b_3 , the corresponding QPSK signals are $s(2n) = b_0 + jb_1$, and $s(2n+1) = b_2 + jb_3$. The coding in STTD is done as shown in Figure 5.14. The first antenna transmits $s(2n)$ and $s(2n+1)$ whereas the second antenna transmits the $-s^*(sn+1)$ and $s^*(2n)$ in the two-symbol period, respectively. Although this coding arrangement is slightly different from that in the original two-branch STBC scheme, the unitary feature of the channel matrix is still retained and as a result a simple linear maximum likelihood detection can still be used. The STTD encoding is optional in UTRAN, but STTD support is mandatory at the UE [9].

In short, STTD is a simple transmit diversity technique that improves the signal quality at the receiver by processing two consecutive samples from two transmit antennas. The advantages of such a system include the following:

- The diversity gain is the same as that of MRC, but only one antenna is needed at the mobile terminal.
- As an open-loop approach, no feedback from the mobile terminal to the base station is required.

5.2.3 Comparison of Smart Antennas and Transmit Diversity

The conventional beamforming concept is based on the far-field condition, which dictates that the received signals at different antenna elements are strongly correlated. When such a condition does not hold, however, the optimal beamforming algorithms such as NLMS should combine both beamforming and diversity combining to produce the optimal antenna weights. In fact, when an adaptive beamforming algorithm is employed, it does not make differentiation between diversity and beamforming. It simply provides such weighting coefficients so that an optimal criterion is met by the output signal, which is typically the maximum signal-to-interference and noise ratio. This is certainly true for the uplink. For the downlink, if there is feedback from the mobile terminals on the transmitted signals from different antennas, an adaptive algorithm can be used at the transmitter. In that case, a unified algorithm can be used to optimize the received signal without differentiating diversity and beamforming. If such a feedback mechanism does not exist, or if the feedback delay is too long, then a decision to employ diversity transmitting or beamforming at the algorithm level must be made.

When a unified optimal adaptation algorithm is used, how the algorithm performs still depends on one important parameter – the spacing between antenna elements. Although an optimum performance can be achieved by using adaptive algorithms, this optimum performance is optimum only on the condition of the given antenna spacing, which determines whether diversity or beamforming dominates. As explained earlier, the potential diversity gain is high if the antenna spacing is large. On the other hand, if the antenna spacing is small, the received signals tend to become strongly correlated, so beamforming becomes dominant. Looking from another angle, the effectiveness of beamforming and diversity depends on the channel condition. For beamforming to be effective, the angular spread should be small, whereas for transmit diversity to perform well, the correlation between different channels should be small. In general, large angular spread and large interelement spacing lead to small channel correlation.

The beamforming antenna is normally aimed at achieving high antenna gain and nulling by shaping the array beam pattern. The transmit diversity, on the other hand, is aimed at maximizing diversity to mitigate the fading effect. If the signal is in deep fade, which can be tens of decibels weaker than the average level, the directional discrimination provided by smart antennas may not help much. On the other hand, the net gain, which can be achieved by introducing higher diversity orders (a greater number of diversity branches) decreases as the diversity order increases, which means that two-branch diversity is quite adequate for most environments and the benefit of going beyond four-branch diversity is rather limited.

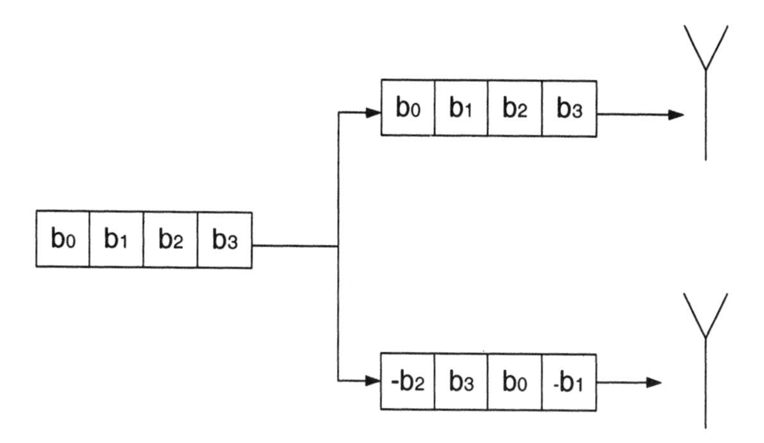

Figure 5.14 The STTD scheme in UTRAN.

From the above discussion, it can be concluded that beamforming is more beneficial to macro cells, whereas transmit diversity is more effective for micro cells. This is because in a micro cell environment, the angular spread can be as large as 360° and the signals from a moderate antenna array are strongly correlated, so beamforming becomes very ineffective. In a macro cell environment, however, the angular spread at the base station tends to be small and the correlations among signals from different antenna elements are weak, so a large interelement spacing is needed to achieve significant diversity gain. Therefore, a very promising scheme in a macro cell environment is to combine adaptive antenna arrays with a two-branch polarization diversity [19], so the benefits of both the beamforming and the transmit diversity can be realized (see Figure 5.15). This configuration has the advantage that, since each antenna element can be dually polarized, only one physical antenna array or panel is needed. A further advantage of the configuration is that it can be employed in both the downlink and the uplink.

5.3 MULTIPLE INPUT MULTIPLE OUTPUT SYSTEMS

In the preceding sections, smart antennas and transmit diversity antennas were discussed as base station features and there is no assumption on the number of antennas at the mobile terminals. With the increasing demand on high data rate services and the advance of DSP chips, a new technology called multiple-input and multiple-output (MIMO) systems that relies on the use of multiple antennas at both the base stations and the mobile terminals is emerging. Current technology developments in wireless and mobile communications indicate a clear trend towards higher data rates, to achieve beyond gigabits per second rates for a nomadic indoor user in wireless LANs and about 10–100 Mbps for highly mobile users in cellular systems. Based on the spatial multiplexing technique, MIMO systems offer a promising way to achieve these rates.

The fundamental principle behind MIMO is the exploitation, rather than the mitigation, of multipath effects in order to achieve very high spectral efficiencies in terms of bits per second per hertz. The spectral efficiency targeted by MIMO is significantly higher than that in conventional wireless systems where multipath is viewed as an impediment rather than an indispensable resource [20–25]. Similar to transmit diversity antennas, MIMO is also based on *diversity* instead of *beamforming*. Therefore, the antenna elements in a MIMO system should be placed in such a way that the correlation between signals arriving at different antenna elements is as small as possible.

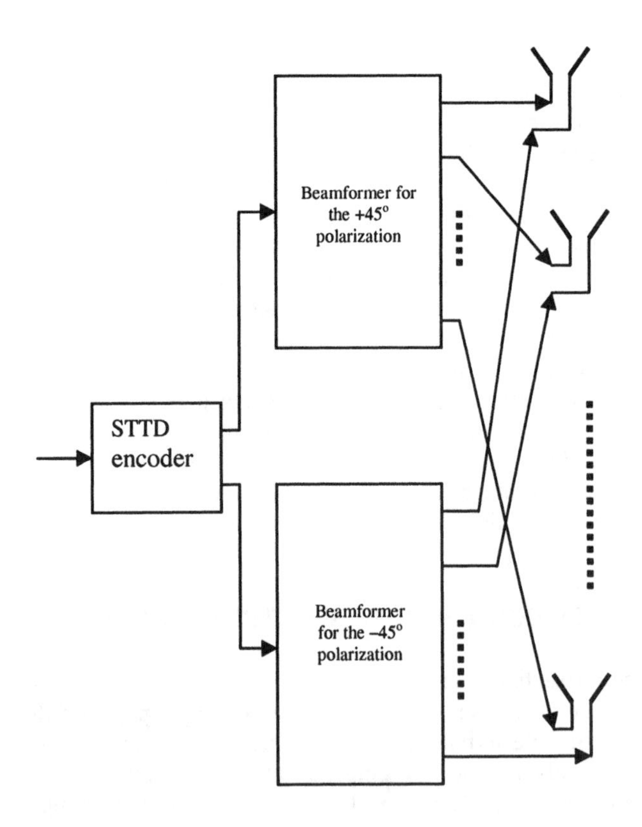

Figure 5.15 Combining beamforming with transmit diversity.

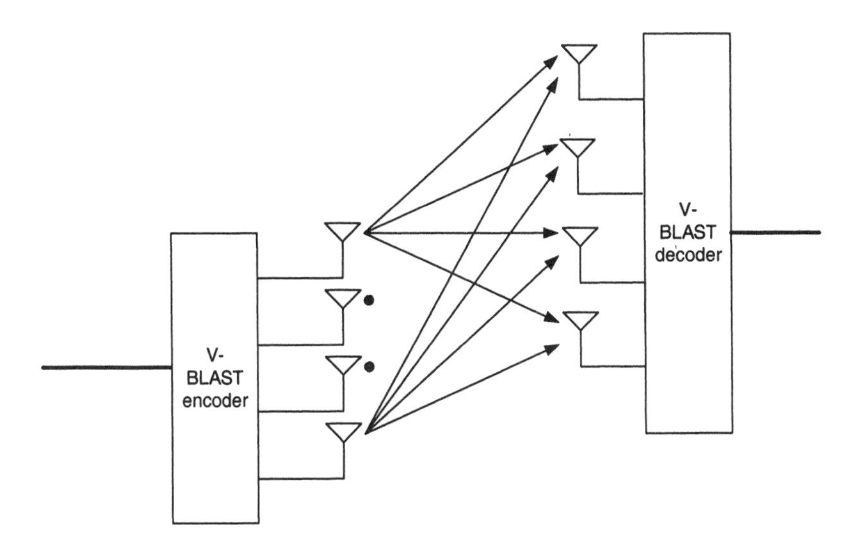

Figure 5.16 Illustration of Bell Lab's MIMO system: BLAST.

Compared with smart antennas and transmit diversity antennas, the main feature of the MIMO system is that, at the transmitter, the input data stream is divided into a multitude of n parallel data streams and these different data streams are transmitted by different antennas. Naturally, this results in an n times increase in the data transmission rate, but no directional gain is achieved. If the transmitting and receiving antennas are far enough or the scattering is rich enough, the employment of m receiving antennas at the receiver makes it possible to separate the n-streams of data by exploiting the uncorrelated nature of the propagation channels, therefore achieving n-fold of that receive data rate. In contrast, in a smart antenna, the same data stream is transmitted from or received by all the antenna elements in the array. In the transmit diversity antenna system, the same data stream is space and time coded and transmitted by different antenna elements.

In a MIMO system, the number of antennas at the receiver m should be no less than that at the transmitter n, and the data rate achievable by the system is mainly dependent on the number of antennas at the transmitter n. Undoubtedly, the quality of n x m radio channels in which a MIMO system operates can be different from each other so the quality of some data streams can be significantly worse than the others. This problem can be solved by using error correction coding to the data streams prior to transmission and by changing the antenna and bit stream association periodically. When employed in a CDMA network, MIMO can lead to very high data rate with only, say, one spreading code, provided that the system is not power-limited. Alternatively, at the same transmission rate, it can be used to save transmission power. As a formal study item in 3GPP (Release 6), MIMO is now a promising capacity enhancement technique for HSPDA discussed in Chapter 4.

One of the best known MIMO schemes is Bell Labs' BLAST (Bell labs Layered Space-Time). A typical BLAST system is depicted in Figure 5.16 [20]. It is seen that the single user's data stream is first split into multiple substreams and these substreams are then transmitted in parallel simultaneously using the antenna array at the transmitter. All the substreams are transmitted in the same frequency band and the same time slot without any spectrum spreading, leading to a very efficient use of the spectrum. As a result, the effective transmission data rate is increased approximately in proportion to the number of transmit antennas. At the receiver, another antenna array is used to receive the transmitted substreams and their multipath reflections. As long as the scattering in the channel is rich enough, all the substreams can be extracted from the received signals.

Although many extraction or separation algorithms have so far been proposed, many challenges still face the research community [26]. For example, how rich should the channel scattering or multipath be to allow all the transmitted substreams to be separated at the receiver under a given signal-to-noise ratio (SNR)? What is the impact of practical channel conditions where the scattering is hardly as perfect as the Rayleigh model on the overall system capacity? In [27], theoretical link capacity is derived when there is a correlation between receive antennas and between transmit antennas. It was found that for a 16 by 16 BLAST system at 10-dB SNR, an interelement spacing of four wavelengths in a linear array with dual polarization allows one to achieve 36 bps/Hz as compared to 42 bps/Hz for the uncorrelated antenna case.

In the past few years, several MIMO system test beds have been built around the world, some of which are described next.

BLAST system The Bell Labs' BLAST system is the first MIMO system reported [20-25]. It realized what were believed to be unprecedented spectral efficiencies, ranging from 20 – 40 bps/Hz. These spectral efficiencies correspond to the payload data rate of 0.5 – 1 Mbps in a 30-kHz channel. For example, Bell Labs has demonstrated a 1.2-Mbps link using eight transmit and twelve receive antennas for indoor applications. By comparison, the efficiencies achieved using traditional wireless modulation techniques range only from around 1 - 5 bps/Hz (mobile cellular) to around 10 – 12 bps/Hz (point-to-point fixed microwave systems). The corresponding data rate achievable in the traditional mobile cellular system (30-kHz bandwidth) is therefore only about 50 Kbps.

AT&T MIMO The research team at AT&T Labs constructed an EDGE-based MIMO system [28]. A downlink data rate of 10 Mbps was achieved in 1.250-MHz bandwidth with four antennas at the base station and four antennas at the mobile terminal. Since it is not very difficult to provide a four-branch diversity with two dual polarization antennas in a laptop type of terminal, this experiment was particularly of commercial interest.

Motorola test bed Motorola has built an OFDM-based MIMO (2x2) test bed which included both a transmit diversity mode and a BLAST type of MIMO mode [29]. The system was operated at 3.7 GHz with 20-MHz bandwidth. Tests were carried out around Schaumburg, IL – a typical urban and suburban environment, and the following preliminary conclusions were reached:

(1) Whenever there is a rich scattering environment, MIMO does deliver a much higher capacity than transmit diversity.
(2) Rich scattering was provided in approximately 30% of the cases or areas. In other words, the transmit diversity delivers better results than MIMO in 70% of the cases.

It should be pointed out that there are several limitations on MIMO. First, if the radio channels are strongly correlated, the data rate that can be realized will be significantly decreased. Second, MIMO requires much-increased signal processing power and power consumption at the receiver, which means a higher cost and lower mobility of terminals. Third, although it is possible to place several antennas with different polarizations at different places at a laptop size terminal, placing several antennas in a compact handset can be very challenging. Fourth, MIMO decoders require the operation of channel estimation but channel estimation errors can degrade the actual performance. Last but not least, MIMO does not offer any directional discrimination to the transmitted signal and therefore it will cause unwanted interference to users in the same or adjacent cells. Owing to these limitations, it is very likely that MIMO technology will find applications mainly in micro and pico cells and nomadic scenarios and for laptop types of mobile terminals.

5.4 CONCLUDING REMARKS

In this chapter, three types of multiple antenna systems have been discussed in the context of future mobile telecommunications networks, namely, smart antennas, transmit diversity antennas, and multiple-input and multiple-output (MIMO) antenna systems. All these three multiple antenna systems can be regarded as different types of space-time processing systems. Starting from a single user data stream for the downlink, some redundant bits are normally introduced to achieve coding gain. In the first case, this coded data stream can be multiplied by different weights and then be fed into different antenna elements, which results in beamforming antennas. The adaptation of the antenna weights according to environment makes the smart antennas the most flexible system and a powerful means of increasing system capacity. Smart antennas for the uplink are based on the same principle but the process is reversed. In the second case, the coded data stream can be space and time coded further and then be fed into different antennas, which is known as transmit diversity. The beauty of the two-branch transmit diversity is that it can increase the receiver performance of the mobile terminal in a fading environment with moderate increase in the digital processing load and without the need for two antennas at the mobile terminals. The disadvantage is that it is only applicable to the downlink. In the third case, the coded data can be partitioned into different substreams and then fed into different antennas with possible individual coding on the subdata streams, which is the concept of MIMO. MIMO has the potential of achieving the highest data rate. Its disadvantage is that the receiver needs to have a multiple antenna system and the demand on signal processing is quite high.

In principle, all these three multiple antenna schemes can be built into the signal processing subsystem of the base station equipment. For actual deployment, however, combining two adaptive digital beamforming antennas with two branch-polarization transmit diversity appears to be a very attractive solution for macro cells. For micro and pico cells, both transmit diversity antennas and MIMO can be used and a choice can be made depending on the terminal capability, cell range, and required data rate.

References

[1] J. C. Liberti and T. S. Rappaport, *Smart Antennas for Wireless Communications*, Upper Saddle River, NJ: Prentice Hall, 1999.

[2] M. J. Feuerstein, "Unlocking the Capacity You Paid for," Seminar one, *CDMA Solution Seminar series*, Metwave Communications Co., 1999.

[3] M. J. Feuerstein, "Smart Antennas for CDMA Cellular Systems," *Proc. of 5th Workshop on Smart Antennas in Wireless Mobile Communications*, Stanford University, July 1998.

[4] J. H. Winters, "Smart Antennas for Wireless Systems," *IEEE Comunications Mag.*, February 1998, pp. 23-27.

[5] R. T. Compton, Jr., *Adaptive Antennas: Concepts and Performance*, Englewood Cliffs, NJ: Prentice-Hall Inc, 1988.

[6] F. Adachi, "Application of Adaptive Antenna Arrays to W-CDMA Mobile Radio," *Proc. of 5th Workshop on Smart Antennas in Wireless Mobile Communications*, Stanford University, July 1998.

[7] Y. J. Guo, S. Vadgama, and Y. Tanaka, "Advanced Base Station Technologies for UTRAN," *Electronics and Communication Engineering Journal*, Vol. 12, No. 3, June 2000, pp. 123-132.

[8] S. Haykin, *Adaptive Filter Theory*, Upper Saddle River, NJ: Prentice Hall, 1996.

[9] 3GPP TS 25.211, Technical Specifications: Radio Access Network, V5.2.0, September 2002.

[10] J. Winters, "Signal Acquisition and Tracking with Adaptive Arrays in the Digital Mobile Radio System IS-54 with Flat Fading," *IEEE Trans. Vehicular Tech.*, Vol. 42, No. 4, November 1993.

[11] A. Molish, "Spatial Channels and Smart Antennas," *IEEE Vehicular Tech. Conference (VTC) - Fall*, Tutorial 3, Amsterdam, the Netherlands, September 1999.

[12] J. G. Proakis, *Digital Communications*, New York: McGraw-Hill, 1995.

[13] C. Passmann and T. Wixforth, "A Calibrated Phased Array Antenna with Polarisation Flexibility for Trunami (II) SDMA Field Trial," *Proceedings of ACTS Mobile Communications Summit* '97, pp. 833 - 838.

[14] C. M. Simmonds and M. A. Beach, "Active Calibration of Adaptive Antenna Arrays for Third Generation Systems," *Proceedings of ACTS Mobile Communications Summit '97*, pp. 870 - 875.

[15] P. E. Morgensen, et al, "Tsunami (II) Stand Alone Testbed," *Proceedings of ACTS Mobile Communications Summit* '96, pp. 517-527.

[16] W. C. Jakes, *Microwave Mobile Communications*, New York: IEEE Press, 1974.

[17] S. M. Alamouti, "A Simple Diversity Technique for Wireless Communications," *IEEE J. Selected Areas in Communi*cations, Vol. 16, No. 8, October 1998, pp. 1451-1458.

[18] V. Tarokh, et al, "Space-Time Block Codes from Orthogonal Designs," *IEEE Trans. Inform. Theory*, Vol. 45, No. 5, July 1999, pp. 1456-1467.

[19] K. I. Pedersen and P. E. Mogensen, "A Simple Downlink Antenna Array Algorithm Based on a Hybrid Scheme of Transmit Diversity and Conventional Beamforming," *Proc. of IEEE VTC 2001 – Spring*, Rhodes, Greece, May 2001.

[20] G. J. Foschini, "Layered Space-Time Architecture for Wireless Communication in a Fading Environment When Using Multiple Antennas," *Bell Labs Tech. Journal*, Vol. 1, No. 2, Autumn 1996, pp. 41-59.

[21] G. J. Foschini and M. J. Gans, "On Limits of Wireless Communications in a Fading Environment When Using Multiple Antennas," *Wireless Personal Commun.*, Vol. 6, No. 3, March 1998, p. 311.

[22] G. D. Golden, et al., "V-BLAST: A High Capacity Space-Time Architecture for the Rich-Scattering Wireless Channel," *5th Workshop on Smart Antennas in Wireless Mobile Communications*, Stanford University, July 1998.

[23] G. D. Golden, et al., "Detection Algorithm and Initial Laboratory Results Using the V-BLAST Space-Time Communication Architecture," *Electronics Letters*, Vol. 35, No. 1, January 1999, pp. 14-15.

[24] G. J. Foschini, et al., "Simplified Processing for Wireless Communication at High Spectral Efficiency," *IEEE J. Select Areas in Communications*, Vol. 17, No. 11, 1999.

[25] D. S. Shiu, et al., "Fading Correlation, and Its Effect on the Capacity of Multielement Antenna Systems," *IEEE Trans. Communi*cations, Vol. 48, No. 3, 2000.

[26] F. C. Zheng and A. G. Burr, "Receiver Design for Orthogonal Space-Time Block Coding for Four Transmit Antennas over Time-Selective Fading Channels," *Proc. of IEEE Globecom 2003*, San Francisco, CA, December 2003.

[27] D. Chizhik et al, "Effect of Antenna Separation on the Capacity of BLAST in Correlated Channels," *IEEE COMMUNICATIONS LETTERS*, Vol. 4, No. 11, November 2000, pp. 337-339.

[28] J. H. Winters, "Smart Antennas for Wireless Systems," VTC2001 Spring Tutorial, *IEEE VTC2001*, Rhodes, Greece, May 2001.

[29] M. D. Batariere, et al., "Wideband MIMO Mobile Impulse Response Measurement at 3.7 GHz," *Proc. of IEEE VTC02-S*, 2002.

APPENDIX 5A: PROOF OF THE CONVERGENCE OF IBS

Defining the steering vector as

$$\mathbf{V} = (1, e^{jkd \sin \vartheta}, e^{j2kd \sin \vartheta}, \cdots, e^{jMkd \sin \vartheta})^T \tag{5A.1}$$

one has

$$\mathbf{X}_n = s_n \mathbf{V} + \mathbf{N}_n \tag{5A.2}$$

where M is the number of antenna elements, k is the wavenumber, θ represents the direction of the wanted signal component, and N_n is a white Gaussian noise vector whose components have zero mean and variance σ^2. Initializing the weights as

$$\mathbf{W}_0 = \mathbf{0} \tag{5A.3}$$

and substituting (A5.2) into (A5.1) gives

$$\begin{aligned} \mathbf{W}_1 &= (1-\alpha)(\mathbf{V} + s_1^* \mathbf{N}_1) \\ \mathbf{W}_2 &= (1-\alpha)[\alpha(\mathbf{V} + s_1^* \mathbf{N}_1) + (\mathbf{V} + s_2^* \mathbf{N}_2)] \\ \mathbf{W}_n &= (1-\alpha^n)\mathbf{V} + \mathbf{N}_{Total}(n) \end{aligned} \tag{5A.4}$$

where

$$\mathbf{N}_{Total}(n) = (1-\alpha)(\alpha^{n-1}\mathbf{N}_1' + \alpha^{n-2}\mathbf{N}_2' + \cdots + \mathbf{N}_n')$$
$$\mathbf{N}_i' = s_i^* \mathbf{N}_i$$

(5A.4) shows that the IBS produces a vector consisting of two parts, the first being proportional to the steering vector and the second being a noise vector $\mathbf{N}_{Total}(n)$. The beamforming weights converge to the following if the noise vector is bounded:

$$\mathbf{W}_{\infty} = \mathbf{V} + \mathbf{N}_{Total}(n \to \infty) \tag{5A.5}$$

The noise power contained in $\mathbf{N}_{Total}(n)$ can be calculated as follows:

$$E\{\mathbf{N}_{Total}(n)^H \mathbf{N}_{Total}(n)\} = M\sigma^2 \frac{(1-\alpha)(1-\alpha^{2n})}{1+\alpha}$$

Letting the number of iterations n approach infinity, one has

$$E\{(\mathbf{N}_{Total}(n \to \infty)^H \mathbf{N}_{Total}(n \to \infty)\} = M\sigma^2 \frac{1-\alpha}{1+\alpha} \tag{5A.6}$$

(5A.5) and (5A.6) show that, in a static channel, the beamforming weights generated by the IBS algorithm converge to the steering vector plus a noise vector, with a signal to noise ratio of

$$SNR_{IBS} = \frac{1+\alpha}{1-\alpha} \tag{5A.7}$$

Chapter 6

Orthogonal Frequency Division Multiplexing Systems

Given the bandwidth, the achievable data rate in a wireless system depends strongly on the radio environment. To be specific, the radio receiver performance tends to degrade when the excessive delay of the channel with respect to the symbol period increases. Although it is possible to deal with large excessive delays with complex receivers, such receivers are normally costly and power-hungry. Therefore, the best strategy is to employ a wireless system that is inherently resilient against the multipath phenomenon; this leads to the concept of orthogonal frequency division multiplexing (OFDM). In an OFDM system, the data stream is divided into M parallel substreams and each of the substreams is transmitted over a different carrier. As a result, the symbol period is effectively extended by M, thus allowing the transmission of much higher data rate.

Currently, OFDM is being used in high data rate wireless systems such as wireless LANs. To increase the data rate of cellular systems, various OFDM schemes are being considered as the air interface for the future mobile communications systems. In this chapter, the operation principle of the OFDM transceivers is introduced first. The practical engineering problems and solutions of the OFDM transmission systems are discussed. Then the theory of an advanced version of OFDM, OFDM/IOTA, is presented. Finally, several OFDM-based proposals for future mobile communications systems are described.

6.1 MULTIPATH AND OFDM

Signal distortion caused by multipath is one of the major challenges in wireless communications. The distortion occurs at the receiver where various objects in the environment reflect part of the transmitted signal energy, and the reflected signals arrive at the receiver with different amplitudes, phases, and time delays. Depending on the relative phase between different reflection paths, individual multipath components can add constructively or destructively at different frequencies. Consequently, the receiver may see some frequencies in the transmitted signal attenuated significantly and others having some relative gain, which is commonly referred to as frequency selective fading. In the time domain, the receiver sees multiple copies of the signal with different time delays. The time difference between different paths often means that different symbols will overlap or smear into each other and create the so-called intersymbol interference (ISI). All modern mobile wireless systems employ some types of techniques to combat the challenges imposed by the multipath channel. These include diversity combining, coding, interleaving, equalization, spread spectrum and orthogonal frequency division multiplexing [1]. Since diversity combining, coding, and interleaving are used in almost all wireless systems, the following discussion is focused on equalization, spread spectrum, and orthogonal frequency division multiplexing.

Equalization is a technique used to overcome the effect of intersymbol interference resulting from time dispersion in the channel and it is primarily used in the time division multiplexing (TDM) systems. Implemented at the receiver, the equalizer attempts to correct the amplitude and phase distortions caused by the channel. Since these distortions change with time, the equalizer must track the changing channel response in order to eliminate the intersymbol interference. To this end, the equalizer is normally fed with a fixed length training sequence at the start of each transmission or periodically, which enables the equalizer to characterize the channel and adapt itself accordingly. Systems based on time division multiple access (TDMA), such as IS-136 and GSM, must use equalizers because the multipath time spread of the channel is typically larger than the symbol period. TDMA systems assign one or more time slots to a user for transmission. There is typically some guard time included between time slots to allow for time tracking errors at the mobile terminal. The use of equalizers adds to the complexity and costs of the TDMA systems, since equalization requires significant amounts of signal processing. The need to transmit a fixed sequence of training bits also adds overhead to the communications, therefore reducing the spectrum efficiency. At this point, it is worth noting that another disadvantage of the TDMA system in a cellular environment is that, due to cochannel interference, not all the carrier frequencies can be used in every cell, and therefore the system needs strict frequency planning, thus resulting in a low frequency-reuse factor of typically 1/6. TDMA systems also have less inherent immunity against multipath fading than spread spectrum systems because they use a much narrower signal bandwidth.

Spread spectrum systems employ frequency diversity, in which the signal is spread over a much larger bandwidth than that of the signal. A wideband signal is more resistant to frequency selective fading than a narrowband signal because only a relatively small portion of the overall bandwidth will experience a fade at any given time. There are two types of spread spectrum systems, frequency hopping spread spectrum (FHSS) and direct sequence code division multiple access (DS-CDMA). In a frequency hopping spread spectrum system, each user's signal is carried by a set of carriers and the set is changed after each time period, usually corresponding to a modulation symbol. By changing frequencies according to a hopping pattern, the losses due to frequency selective fading are reduced. In a DS-CDMA system, each user's data is spread across the whole bandwidth by multiplying the information-bearing data stream with a spreading sequence code of a higher rate prior to transmission. A major advantage of the DS-CDMA systems such as WCDMA and CDMA2000 1x is that the Rake receiver can be employed to take advantage of the multipath phenomenon by picking up the multipath components of a signal individually at different times and combining them in a constructive manner, thus exploiting time diversity. A further advantage of the DS-CDMA is that, unlike frequency hopping, all the bandwidth can be reused by all the users in all the cells in a cellular environment by employing different spreading sequences and no frequency planning for different cells is needed.

However, there are some problems in DS-CDMA systems. One is that different code sequences used for spreading different user data are not truly orthogonal in the presence of multipath delay spread. This results in interference between users within a cell, which is referred to as multiple access interference (MAI), and MAI ultimately limits the capacity of the cell. In fact, more than two thirds of the interference in a DS-CDMA sector typically comes from the users in the same sector. Another problem in the DS-CDMA system is that the complexity of Rake receivers increases with the expected excessive delay spread relative to the symbol period. Generally speaking, a practical Rake receiver with limited complexity and power consumption performs well if the delays between different components carrying significant energy are much less than the symbol period. When a very high data rate is needed, however, the relative delay becomes greater and the bit error rate in the receiver tends to increase, eventually resulting in communications failure. Nevertheless, DS-CDMA systems do possess a performance advantage of interference averaging, which enables a system to be engineered for the average interference experienced rather than the worst-case interference, thus allowing universal frequency reuse in a cellular environment. This increases overall system capacity.

In contrast, the OFDM scheme is inherently robust against the multipath problem in very high data rate communications. Its essence is to transmit a large number of parallel low-speed data streams instead of one single high-speed data stream, thus reducing the relative delay spread of the signal with respect to the symbol period. In OFDM, a number of equally spaced tones (carrier frequencies) are chosen and each tone carries a portion of a user's information. OFDM can be viewed as a form of frequency division multiplexing (FDM), but it has an important special property that the signals carried by different tones are orthogonal to every other. FDM normally requires some frequency guard bands between the adjacent frequencies so that they do not interfere with each other. However, OFDM allows the spectrums of different tones to overlap, and the orthogonality between them guarantees that they do not interfere with each other. By allowing the tones to overlap, the overall bandwidth required to achieve given throughput is reduced. When OFDM is employed for multiple access applications, individual tones or groups of tones can be assigned to different users to enable multiple users to share a given bandwidth, thus yielding a system called orthogonal frequency division multiple access (OFDMA). In an OFDMA system, each user can be assigned a predetermined number of tones when they have information to send, and the number of tones can be determined according to the required data rate and loading conditions in the cell.

OFDM relies on multiple narrowband subcarriers to carry all the data. In multipath environments, the subcarriers at frequencies that suffer from more fading will be received with lower signal strength, thus resulting in an increased error rate for the bits transmitted on these weakened subcarriers. In most practical environments, fortunately, this frequency fading only affects a small number of subcarriers and therefore it only increases the error rate on a portion of the transmitted data stream. As a result, these erroneous bits can be recovered by interleaving and error correction coding. To improve its immunity against frequency selective fading, OFDMA can be further combined with frequency hopping to exploit the benefits of frequency diversity and interference averaging associated with CDMA.

6.2 BASIC OFDM TRANSMITTERS AND RECEIVERS

In an orthogonal frequency division multiplexing (OFDM) system, a set of equally spaced carriers are selected with each carrying a portion of the whole transmitted signal, thus resulting in a parallel transmission of different bits at different frequencies. This modulation scheme is often termed "multicarrier" or "multitone," as opposed to the conventional "single carrier" schemes. Each individual carrier, commonly called a subcarrier, transmits information by modulating the phase and possibly the amplitude of the subcarrier over the symbol duration. That is, each subcarrier employs a modulation scheme to convey information just as conventional single carrier systems. Since the user data is separated and transmitted over the air in a large number of subdata streams carried by subcarriers, the data rate on each subcarrier is significantly reduced. The spacing between these subcarriers in the frequency domain is selected to be the inverse of the symbol duration so that each subcarrier is orthogonal or noninterfering. This frequency spacing is the smallest that can be used without creating interference [1].

Figure 6.1 shows the schematics of an OFDM transmitter with QAM modulation. It is seen that the input data stream is first encoded by a QAM modulator and then separated into M parallel substreams by virtue of a serial-to-parallel converter. Each substream of data is modulated by its individual carrier and the summed signal is transmitted by the RF part of the transmitter after up-conversion (not shown). For the receiver, the whole process is reversed. The signal is down-converted and separated by the filters, and the M parallel symbol streams are converted to a single faster symbol stream and decoded by the demodulator, which is shown in Figure 6.2. From Figures 6.1 and 6.2, it might appear that OFDM systems must modulate and demodulate each subcarrier individually, and therefore are necessarily complicated. In fact, in a practical OFDM system, the fast Fourier transform (FFT) chips are employed for modulating and demodulating those parallel subcarriers as a group rather than individually, thus resulting in a low implementation cost.

Figure 6.3 shows the details of a practical OFDM transmitter based on inverse FFT (IFFT). After QAM modulation, the serial symbol stream is converted into M parallel streams by virtue of a serial-to-parallel converter. These M streams are then modulated onto M subcarriers by using size N ($N>M$) inverse FFT (IFFT). The N outputs of the inverse FFT are then serialized to form a data stream that can then be modulated by a single carrier normally at the central frequency of the given band. Note that the N-point inverse FFT can modulate up to N subcarriers. When M is less than N, the remaining N-M subcarriers can be suppressed in the output stream by modulating them with zero amplitude. As an example, the IEEE802.11a standard for wireless LAN specifies that 52 ($M = 52$) out of 64 ($N = 64$) possible subcarriers are modulated by the transmitter [2].

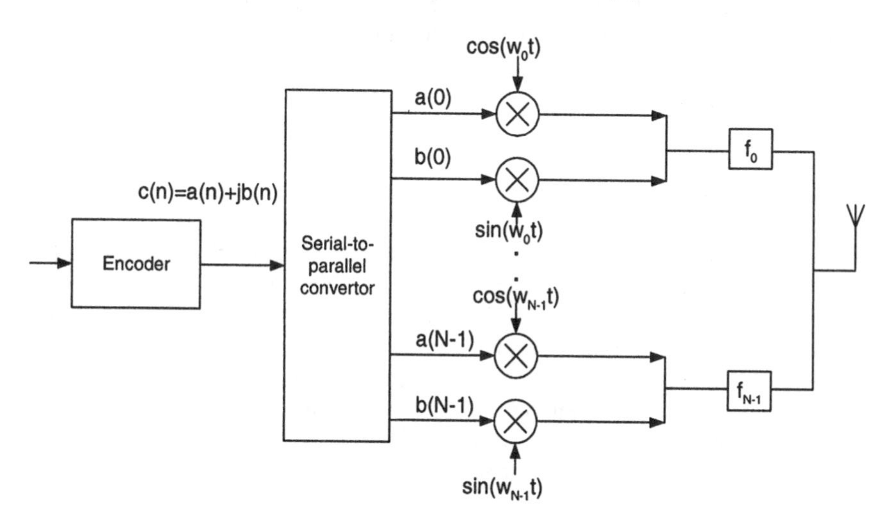

Figure 6.1 Block diagram of a simple OFDM transmitter.

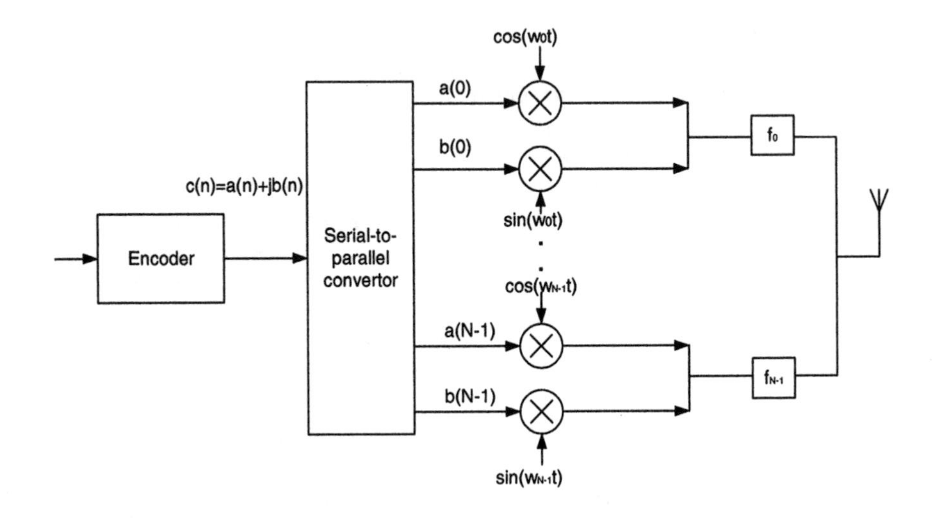

Figure 6.2 Block diagram of a simple OFDM receiver.

Figure 6.4 illustrates the process at the receiver. The received data is split into N parallel streams that are processed with a size N FFT. The size N FFT efficiently implements a bank of filters each matched to the N possible subcarriers. The FFT output is then serialized into a single stream of symbols for decoding. Note that when M is less than N, the receiver only serializes the M subcarriers with data.

6.3 PRACTICAL ISSUES

While the overall design of an OFDM system is simplified because of its robustness against multipath and the use of FFT chips, there are several practical issues that must be addressed in order to realize these advantages. In fact, several major practical problems have hindered its widespread applications in many areas in the past. For instance, the problem of large peak-to-average power ratio (PAPR) played a major role in preventing OFDM from being adopted as the air interface for the European wireless LAN standards, HIPERLAN, and, to some extent, for UTRAN. Fortunately, recent progress in signal processing and IC technology has made these issues a lesser problem.

In the following, four major engineering problems that have a strong impact on the performance of OFDM systems–PAPR, guard band, frequency offset, and phase noise–are presented and possible solutions are discussed [1, 3].

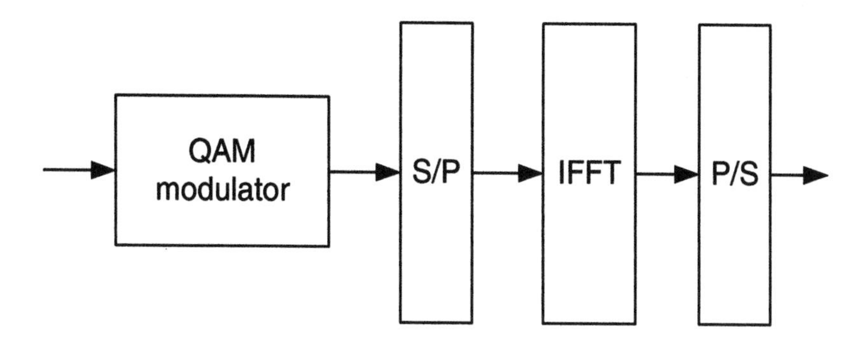

Figure 6.3 Block diagram of an OFDM transmitter based on IFFT.

6.3.1 Peak-to-Average Power Ratio

A major challenge in the design of OFDM systems is to deal with the large dynamic range of the signal power. The range of the signal power is normally measured by the peak-to-average power ratio (PAPR). For OFDM signals, the maximum signal power can be significantly greater than the minimum and average power, thus resulting in very high PAPR. Large PAPR is inherent to multi-carrier modulations since each subcarrier carries essentially independent data streams. As a result, signals carried by different subcarriers can add constructively and destructively in the time domain and this creates the potential for a large variation in the instant signal power. In other words, it is possible for the data sequence to make all the subcarriers align constructively and sum to a very large signal. It is also possible for the data sequence to make all the subcarriers align destructively and sum to a very small signal. This wide variation creates a number of problems mainly for the transmitter because it needs to accommodate a large range of signal power with a minimum of distortion.

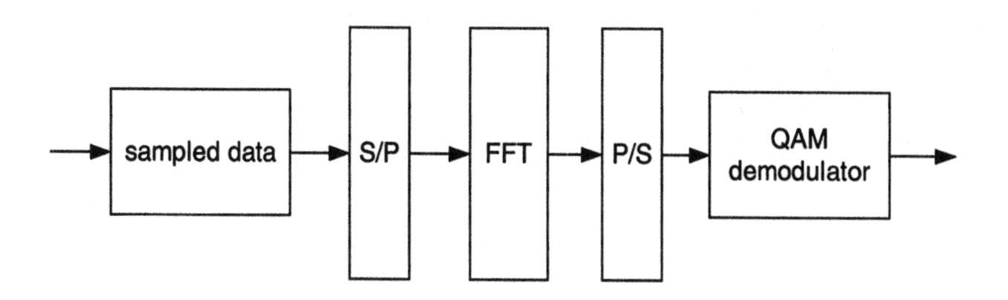

Figure 6.4 Block diagram of an OFDM receiver based on FFT.

The large dynamic range of OFDM systems presents a particular challenge to the power amplifier (PA) design. Practical power amplifiers have both linear and nonlinear regions (see Chapter 2). The nonlinear region corresponds to large output powers, and it distorts the signal and produces out-of-band radiations. To reduce the amount of distortion and out-of-band energy generated by the transmitter, OFDM and other modulations need to operate as much as possible in the linear region. With its inherently large dynamic range, this means that OFDM must keep its average power well below the non-linear region in order to accommodate the signal power peaks. As discussed in Chapter 2, lowering the average power reduces the efficiency of the power amplifier. On the other hand, since the occurrence of large signal powers is rare, this sacrifice is made to the majority of the signal but for the sake of infrequent events. Therefore, in the design of practical OFDM systems, it is necessary to make a careful trade-off between distortion and PA efficiency. The chosen average input power level must be high enough to ensure adequate output power but not too high to introduce too much nonlinear distortion.

To examine this trade-off further, consider the IEEE802.11a version of OFDM that employs 52 subcarriers. In theory, all 52 subcarriers could add constructively and this would yield a peak power of 10log(52) = 17.2 dB above the average power. However, this is an extremely rare event from the viewpoint of statistics. Instead, most simulations show that for real PAs, accommodating a peak that is 3 to 6 dB above average is sufficient. In other words, the distortions caused by the peaks above this range are infrequent enough to allow for low average error rates. Admittedly, the exact margin needed to accommodate the signal peaks is highly dependent on the PA characteristics and the specific system in question.

The large PAPR of the OFDM signal contributes to the complexity and cost of the transceiver system. In the last decade, a substantial amount of research on the subject has been carried out and some elegant methods of alleviating the PAPR problem have been found. One interesting approach is to reduce PAPR by constraining the modulation sequences of the subcarriers. In other words, the subcarriers are no longer independent but rather have a defined phase and amplitude relationship that is chosen to keep the PAPR small. Along these lines, special block codes have been developed that generate low PAPRs. Using these special block codes for the subcarrier modulation instead of allowing the data to modulate the subcarrier freely can significantly reduce the peak-to-average ratio. In addition to reducing the PAPR, these block codes add an additional layer of error correcting capability to the system. Another simple method for handling PAPR is to limit the peak signals either by clipping or by replacing peaks with a smooth but lower amplitude pulse. Since this modifies the signal artificially, it does increase the distortion to some degree. However, if done in a controlled fashion, the distortion introduced by this method should be limited.

It is expected that future mobile communications systems will be data-centered and therefore they will be designed to deal with packets efficiently. In such a system, the automatic retransmission request (ARQ) scheme is normally employed and the receiver can request a retransmission of any packet with uncorrectable errors. Consequently, the PAPR can be effectively controlled by applying scramble sequences to the data prior to modulation, which randomizes the occurrence of large power peaks for each packet transmission. Admittedly, this approach does not prevent large peaks and there will still be occasions when the transmitter introduces significant distortion due to a large peak power in the packet. When the distortion is too severe for the receiver to decode the packet correctly, the latter will request a retransmission. When the data is retransmitted, however, a different scramble sequence is employed. If the first scramble sequence has caused a large power peak, it is very unlikely that the second sequence will do the same despite the fact that it contains the same data sequence. This technique has been used in IEEE 802.11a networks to mitigate the large PAPR problem. The disadvantage of this technique is that it does impact the network throughput due to the fact that some of the data sequences must be transmitted more than once.

6.3.2 Guard Interval

OFDM systems are designed to mitigate the intersymbol interference (ISI) problem caused by multipath delay. Owing to practical limitations on the number of subcarriers, however, the ISI effect cannot be removed completely. To avoid one symbol smearing into the next, a "guard interval" can be added to each symbol, during which some cyclic OFDM extension is inserted. A common way of doing the insertion is to insert the cyclic prefix, in which the last part of the OFDM symbol is placed at the beginning, which is now explained further.

If one assumes that the bandwidth of the OFDM signal is finite, then the Fourier theory dictates that the duration of the corresponding time domain signal is infinite. The underlying assumption upon invoking the IFFT operation for modulation is that although N frequency domain samples produce N time-domain samples, the time-domain signal is periodical. In practice, however, it is sufficient to repeat the time-domain signal periodically for the duration of the channel's memory, that is, the duration of the meaningful channel impulse response. Hence, for the transmission of the OFDM signal, each OFDM symbol that is defined as the set of coefficients at the output of the IFFT can be extended in a cyclic manner to fill the guard interval. Since no new information is conveyed in the cyclic extension, the receiver can ignore the guard interval and still be able to separate and decode the subcarriers. If the guard interval is designed to be longer than any smearing due to multipath, the receiver is able to eliminate ISI distortion by discarding the unneeded guard interval occupied by the cyclic prefix. Hence, ISI is removed with virtually no added receiver complexity.

It is important to note that discarding the guard interval does have an impact on the signal-to-noise ratio, as it reduces the amount of energy available at the receiver for decoding. If M is the original length of a block, and the channel's response is of length L, the cyclically extended symbol is of length $M + L - 1$. At the OFDM receiver, only M samples are processed, and $L - 1$ samples are discarded, thus resulting in a loss in signal-to-noise ratio (SNR) given by

$$SNR_{LOSS} = 10\log(\frac{M+L-1}{M}) \tag{6.1}$$

Also, the cyclic extension reduces the data rate as no new information is contained in the added guard interval. Thus, a good system design will make the guard interval as short as possible while maintaining a sufficient multipath protection.

6.3.3 Frequency Offset

Another challenge inherent in the OFDM system design is to remove frequency offset. Frequency offset can occur when the voltage-controlled oscillator (VCO) at the receiver is not oscillating at exactly the same carrier frequency as the VCO in the transmitter. For the receiver, this offset between the two VCOs is seen as a frequency offset in the signal and it can lead to an increase in the error rate [4]. While this problem is a general one for all modulations, OFDM is particularly sensitive to frequency offsets due to the use of FFT. When there is no frequency offset, the matched filters line up perfectly with the received signal and there is no interference between the subcarriers at the matched filter output. When there is frequency offset, however, the received signal is shifted in frequency and, as a result, the matched filters are offset from the received signal. Consequently, energy from adjacent subcarriers will become mixed at the output of each matched filter. In other words, from the receiver's viewpoint, the subcarriers are no longer orthogonal. This leads to the intercarrier interference (ICI) as adjacent subcarriers interfere with other. If ICI is neglected in the system design, the error rate at the OFDM receiver can increase rapidly with the frequency offset.

There are several techniques for estimating and removing frequency offset. For example, in packet-based systems such as IEEE 802.11a, a training sequence is usually placed at the beginning of the packet. This training sequence is specifically designed to aid the receiver in estimating the frequency offset between the transmitter and receiver. Once the offset is known, it can be removed by adjusting the frequency of the VCO either in analog or digital hardware. Alternatively, nondata-aided techniques can be used to estimate the frequency offset in OFDM systems in an adaptive manner [1].

6.3.4 Phase Noise

In addition to the frequency offset discussed in Section 6.3.3, the frequency generated by a practical VCO tends to jitter over time. To the receiver, this frequency variation appears as noise. The oscillator noise stems from oscillator inaccuracies in both the transmitter and the receiver, and it manifests itself in the baseband as additional phase and amplitude noises added to the received samples. Since the effect of the amplitude noise on the data samples can be neglected, this impairment is commonly referred to as phase noise. In many cases, the frequency variation is slow relative to the signal and the receiver can track and remove the resulting phase noise by using a phase-lock loop (PLL).

For OFDM systems, the design of the PLL can be simplified by inserting training data into the symbol stream. The use of training symbols is common in single carrier systems as well. With OFDM, however, there are some subtle differences. Unlike single carrier systems where training symbols are inserted periodically in time, every OFDM symbol contains a few subcarriers that are modulated with the known training data. These special subcarriers are usually referred to as pilot tones. In the IEEE 802.11a standard, for instance, 4 of the 52 subcarriers are used as pilot tones and the pilot tones are modulated with a binary phase-shift keying (BPSK) sequence that is known to the receiver. Since the modulation sequence is known, these pilot tones can be used to track phase variations due to VCO jitter, thus allowing IEEE 802.11a receivers to remove a majority of the phase noise seen at the receiver.

6.4 OFDM/IOTA

The QAM modulation scheme is normally used in the classical OFDM systems. For convenience, these systems are referred to as OFDM/QAM in the following. Although OFDM/QAM systems do offer great robustness against multipath, they can be potentially improved further by replacing the QAM modulation with the offset QAM (OQAM) modulation and changing the pulse shaping function to IOTA, where IOTA stands for isotropic orthogonal transform algorithm, thus resulting in the OFDM/IOTA scheme [5, 6].

The pulse shaping function used in OFDM/QAM signal to modulate each subcarrier is localized in the time domain, but it extends to infinity in the frequency domain. To be specific, although a rectangular waveform is used in the time domain, its Fourier transform is a $\sin(x)/x$ function whose sidelobes decay very slowly in the frequency domain. This is not a problem if there is no frequency distortion in the signal path, as the signals carried by different frequencies are orthogonal at the chosen carrier frequencies. If the frequency domain has distortions that include frequency offset and phase noise at the receiver, the Doppler effect and nonlinear distortion at the power amplifier occur in the signal path; however, the bit error rates at the receiver can increase significantly. In theory, as explained in Section 6.3, the frequency offset observed at the receiver can be estimated and compensated, but this would result in system complexity and loss of spectrum efficiency due to the use of training sequences. Another problem in the OFDM/QAM is that, as discussed in Section 6.3, a guard time (cyclic prefix) is needed to reduce intersymbol interference and thus protecting the signal from severe multipath delays. Since no information is transmitted during the guardtime, the spectral efficiency is reduced. Fortunately, these two problems can be alleviated effectively by employing the OFDM/IOTA scheme.

In order to localize the pulse-shaping function not only in the time domain but also in the frequency domain, one needs to select such a pulse-shaping function that both its Fourier transform and the function itself decrease fast with frequency and time, say, faster than $|f|^{-3/2}$ and $|t|^{-3/2}$, respectively. On the other hand, it is required that the OFDM/IOTA signals be orthogonal in both the time and the frequency domains. In other words, orthogonality among multicarrier symbols is needed so that no error would be observed at the receiver in a noiseless and distortionless channel. Unfortunately, such a pulse-shaping function does not exist in the traditional sense. However, functions that guarantee orthogonality only in a real domain do exist. Mathematically, this means that one needs to change the definition of the inner product in Hilbert space from

$$\langle x|y\rangle = \int x(t)y^*(t)dt \tag{6.2}$$

to

$$\langle x|y\rangle = \mathrm{Re}\int x(t)y^*(t)dt \tag{6.3}$$

where * denotes complex conjugate and Re represents the operation of taking only the real part. As shown in the following, the engineering implication of (6.3) is that the complex data stream (c_{mn}) must be separated into its two real components: the real part (a_{mn}) and the imaginary part (b_{mn}) (see Figure 6.5), and the imaginary part must be modulated with a half-symbol duration ($T/2$) shifted version of the modulation filter (pulse-shaping function) for the real part, where T represents the symbol duration. In other words, the offset QAM (OQAM) modulation must be used to replace the QAM modulation. It should be pointed out that OQAM is an old concept and it was initially introduced to communications system in order to reduce the envelope variations of the transmitted signal and, in particular, to prevent the envelope of the transmitted signal passing through zero during transitions between symbols, as this would require the power amplifier to maintain linearity over a wide amplitude range.

The classical OFDM signal without a cyclic prefix can be written as:

$$s(t) = \sum_{n=-\infty}^{n=+\infty} \sum_{m=0}^{m=N_u-1} c_{mn} e^{2i\pi m\Delta ft} g(t-nT) \tag{6.4}$$

where $g(t)$ is a rectangular pulse-shaping filter. The basis function is given by

$$x_{m,n}(t) = e^{2i\pi m\Delta ft} g(t-nT)\,. \tag{6.5}$$

For an OFDM/OQAM signal, one has

$$s(t) = \sum_{n=-\infty}^{n=+\infty} \sum_{m=0}^{m=N_u-1} a_{mn} i^{m+n} e^{i\pi mt/T} g(t-nT) + ib_{mn} i^{m+n} e^{i\pi mt/T} g\left(t+T/2-nT\right) \tag{6.6}$$

Note that

$$\Delta f \cdot T = 1/2 \tag{6.7}$$

and a multiplier i^{m+n} have been introduced to (6.6) to ensure orthogonality under (6.3). The basis function in the OFDM/OQAM scheme that is used for representation of any signal can be defined as

$$x_{m,n}(t) = e^{i\pi mt/T} i^{m+n} g(t-nT) \tag{6.8}$$

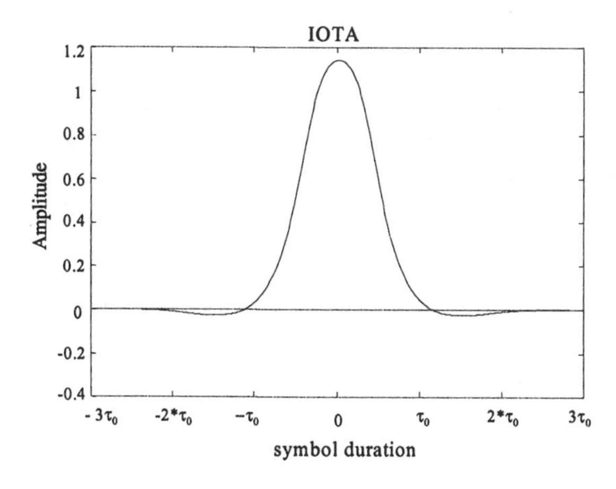

Figure 6.5 Illustration of the IOTA function.

OFDM/IOTA is a special case of the OFDM/OQAM. For OFDM/IOTA, the basis function is defined as

$$\Im_{m,n}(t) = i^{m+n} e^{\sqrt{2} i \pi m t / T} \Im(t - nT / \sqrt{2}) \tag{6.9}$$

where the IOTA filtering function $\Im(t)$ is given by

$$\Im(t) = (2)^{1/4} F^{-1} OFOe^{-\pi t^2} \tag{6.10}$$

In (6.10), F denotes the Fourier transform operation and O denotes the orthogonalization operator defined as

$$Ox(t) = \frac{2^{1/4} x(t)}{\sqrt{\sum_k \left\| x(t - k / \sqrt{2}) \right\|^2}} \tag{6.11}$$

where $\| \ \|$ represent the norm in Hilbert space. The orthogonality of the IOTA function is expressed as:

$$\mathrm{Re}\left(\int_{\Re} \Im_{m,n}(t).\Im^*_{m',n'}(t)dt \right) = \delta_{m,m'}\delta_{n,n'} \tag{6.12}$$

Figures 6.5 and 6.6 show the IOTA function and its Fourier transform, respectively. It is seen that the IOTA filtering function $\Im(t)$ has the following properties:

- It is identical to its Fourier Transform, so the OFDM/IOTA signal is affected similarly by the time and frequency spreading due to propagation conditions.
- The time-frequency localization is quasi-optimal as the IOTA function is similar to the Gaussian function (optimally localized).

Under the OFDM/IOTA scheme, any signal can be expressed as

$$s(t) = \sum_{m,n} a_{m,n} \Im_{mn}(t) \tag{6.13}$$

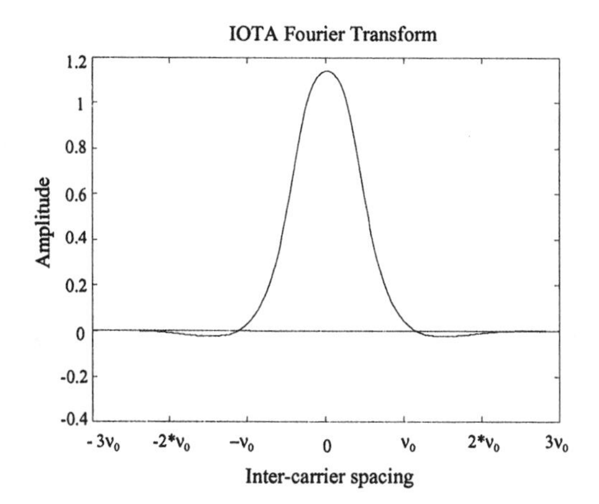

Figure 6.6 The Fourier transform of the IOTA function.

The Fourier transform of (6.13) is given by

$$S(f) = \sum_{m,n} (-1)^{mn} a_{m,n} e^{i(m+n)\pi/2} e^{-\sqrt{2}i\pi nf} \Im(f - m/\sqrt{2}) \quad (6.14)$$

where

$$a_{m,n} = \Re e \int s(t) \Im^*_{m,n}(t) dt \quad (6.15)$$

(6.14) forms the theoretical basis of the OFDM/IOTA scheme. For implementation, the multicarrier part of the modulation and demodulation is done by virtue of the FFT algorithm, whereas filtering is done by virtue of polyphase filter banks [7]. As an illustration, the block diagrams of the OFDM/IOTA transmitter and receiver are shown in Figures 6.7 and 6.8, respectively.

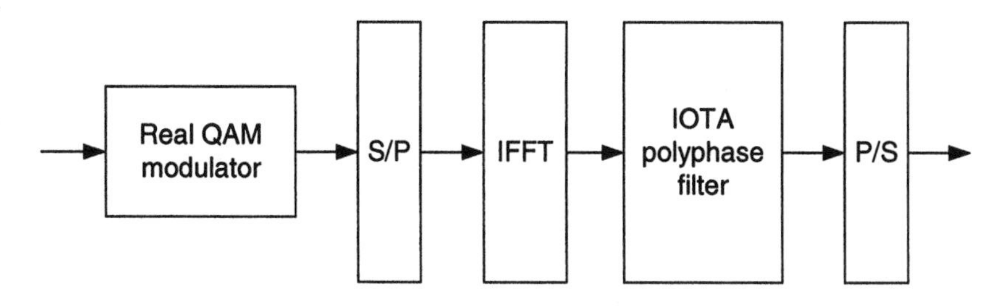

Figure 6.7 OFDM/IOTA transmitter.

The frequency property of the IOTA function makes the OFDM/IOTA much more resilient against frequency distortion because of the decaying tails. Similarly, the fast vanishing nature of the IOTA in the time domain means that the guard interval is not needed or can be reduced. For these reasons, the OFDM/IOTA scheme is being studied in 3GPP for the evolution of the air interface in UTRAN [8-10].

6.5 OFDMA FOR MOBILE RADIO ACCESS SYSTEMS

Owing to its advantages over TDMA and CDMA, OFDMA is being studied as a multiple access technology for future wireless and mobile communications. Generally speaking, the following features make OFDMA well suited to the mobile radio environment:

- **Great immunity against multipath distortion**
  The use of several parallel subcarriers in OFDM results in longer symbol duration, which makes the signal inherently robust to time dispersion. Furthermore, a guard time can be added to combat the ISI further without increasing receiver complexity.
- **High spectral efficiency**
  OFDM is constructed with fully orthogonal carriers, thus resulting in tight frequency separation and high spectral efficiency.
- **Low-cost baseband processing**
  Since the baseband processing is mainly executed by standard FFT chips, the cost of the transceiver is low.

Currently, there are three types of industrial groups working on the OFDM based systems for future wireless and mobile communications: the IEEE 802.x, individual companies such as NTT DoCoMo and Flarion, and 3GPP.

6.5.1 IEEE 802 Systems

Following the widespread adoption of the wireless LAN (WLAN) standard-set IEEE 802.11x, two new industrial groups within IEEE have been formed to apply OFDM technology to other wireless communications areas and to create new standards [2]. The first is the relatively mature IEEE 802.16a working group whose scope is mainly wireless MAN (WMAN) [11]. The main application of the wireless MAN is to establish point to multipoint links between WLANs and operators' IP networks. One substandard defined by IEEE 802.16a is the wireless MAN-OFDMA, which employs the orthogonal frequency division multiple access (OFDMA) using a 2,048-point FFT and the IEEE 802.16 medium access control (MAC) protocol. In this system, multiple access is provided by addressing a subset of the multiple carriers to individual users. The IEEE 802.16a standard was approved in January 2003. Compared with WLAN, the wireless MAN system offers higher QoS, greater user capacity, higher data rates up to 70 Mbps as opposed to 54 Mbps in the former, and longer communications range. As with other advanced wireless data systems such as HSDPA in UTRAN and CDMA2000 1x EV DV, an adaptive modulation scheme is employed in wireless MAN, which includes QAM, 16QAM, and 64QAM.

The scope of the newly formed IEEE 802.20 working group is to create a new standard for future mobile communications, which is referred to as mobile broadband wireless access (MBWA) [12]. No standard for MBWA has been created yet, but it will be a cellular-oriented technology probably based on OFDM. Currently, the debate within IEEE 802.20 is mainly on the issue of frequency division duplex (FDD) against time division duplex (TDD). The disadvantages of TDD include problems associated with frequent transmitter turn-on and turn-off and cell-range dependent guard time. The argument against FDD is the inefficient symmetrical resource utilization to accommodate an asymmetric traffic flow, although this is highly dependent on the future user traffic pattern.

A strong candidate for MBWA is the FLASH-OFDM developed by Flarion, a U.S. company specializing in OFDM cellular systems [13]. In order to accommodate cellular structure and to control intercell interference, frequency hopping has been employed in Flarion's FLASH-OFDM, in which users in different cells are allocated with subgroups of carriers and these carriers follow predefined hopping patterns. Intercell interference is avoided by employing orthogonal hopping patterns in adjacent cells. A further advantage of employing frequency hopping is that frequency diversity is obtained. For users within the same cell, different carriers or tones are employed so they do not interfere with each other. However, the disadvantage of the scheme is that, since only a subgroup of the carriers are used for carrying information in each cell, the instantaneous frequency reuse factor is effectively reduced compared with DS-CDMA. In a DS-CDMA system, the frequency reuse factor is 1 but the intracell and intercell interference is suppressed by spectrum spreading instead of being removed. In the frequency-hopping OFDM, the intercell and intracell interference is removed, but the frequency reuse for each subcarrier is limited, although one may argue that the global frequency reuse in terms of the given band is still 1.

In FLASH-OFDM, information-carrying symbols based on QAM modulation schemes are loaded on each tone for each OFDM symbol duration. Fast hopping across all tones is performed in a pseudorandom predetermined pattern. A user that is assigned one tone does not transmit on the same tone every symbol, but uses a hopping pattern to jump to a different tone in every symbol duration. Different base stations use different hopping patterns and each uses the entire available spectrum over time.

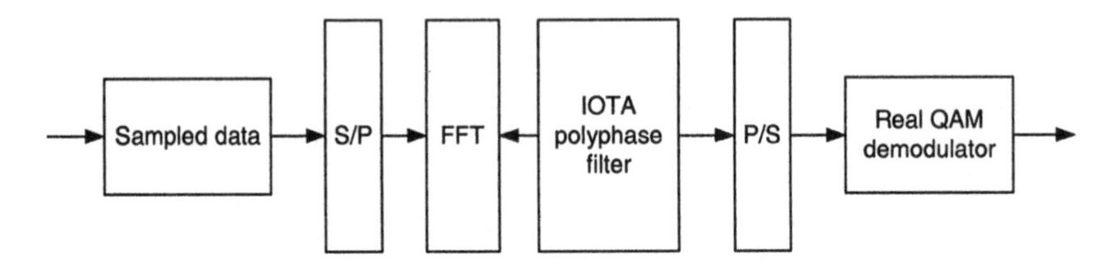

Figure 6.8 OFDM/IOTA receiver.

6.5.2 NTT DoCoMo's 4G System

The Japanese operator NTT DoCoMo is developing a new air interface for fourth generation mobile communications systems [14, 15]. The essence of the scheme is to combine direct sequence spread spectrum technology with the OFDM technology in such a manner that OFDM technology can be used effectively in both cellular and isolated hot-spot environments.

In the downlink, the modulation scheme used by NTT DoCoMo is called variable spreading factor-orthogonal frequency and code division multiplexing, (VSF-OFCDM). VSF means that different spreading sequences are used to reduce intercell and intracell interference and the variable spreading factor is employed to accommodate different data rates. OFCDM is developed from the multicarrier CDMA scheme [16]. In OFCDM, the spreading is applied to the data symbol in both the time domain and the frequency domain on top of OFDM. To be specific, the data carried by each tone is spread by a spreading sequence and for low data rate services the user data can be further spread in the frequency domain to make use of the available subcarriers. The overall spreading factor is the multiplication of the spreading factors in two domains. The priority spreading is done in the time domain and the frequency-domain spreading is secondary. For low data rate signals, the combination of the time-domain and frequency-domain spreading can result in a large spreading factor. The spreading factor in the time domain is changed according to the data rate and the Doppler effect, and a high data rate and a large Doppler effect would lead to the use of short spreading sequences. In the frequency-domain, the spreading factor is changed according to the data rate and the delay spread, as a low data rate would allow a higher frequency spreading and a large delay spread would require the use of more subcarriers. The maximum spreading factor in the time and the frequency domains is 16, thus resulting in the maximum overall spreading factor of 256. In an isolated hot-spot environment, the spreading factors can be chosen as 1 and then it becomes an OFDMA system.

In the uplink, the direct sequence CDMA with variable spreading and chip repetition factors, which is referred to as VSCRF-CDMA, is employed. The idea is to use repetitive sequences as spreading codes so that the spectrum of the latter becomes comb-shaped. By assigning a specific phase vector to each user, the resulting signal becomes orthogonal in the frequency domain. In a typical cellular environment, the chip repetition factor (CRF) is chosen as 1, and a cell specific scrambling code is used, which leads to a conventional DS-CDMA system with 100% cell frequency reuse. In hot-spot scenarios where the intercell interference is not a serious concern, however, the chip repetition is employed to reduce intracell interference and increase cell capacity in the uplink. The block diagrams of NTT DoCoMo's base station and mobile terminals are shown in Figures 6.9 and 6.10, respectively.

6.5.3 OFDM/IOTA for HSDPA in UTRAN

In an effort to increase the system capacity for data-centric applications, 3GPP is studying the possibility of introducing OFDM technology to UTRAN as the next evolution step. To this end, several companies are studying OFDM/IOTA as an alternative scheme to OFDM/QAM. One promising scenario for applying OFDM/IOTA is to introduce a new air interface for the high-speed downlink packet access (HSDPA) in order to increase the throughput in the downlink. To be specific, it has been proposed to allocate a separate 5-MHz bandwidth for UTRAN to a new OFDM HS-PDSCH channel. As shown in Figure 6.11, the node B will need to support two air interfaces, WCDMA and OFDM for HSDPA.

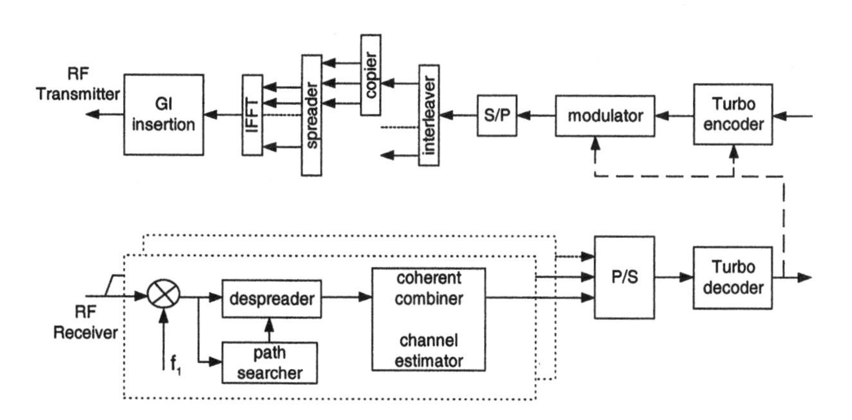

Figure 6.9 A block diagram of NTT DoCoMo's 4G base station.

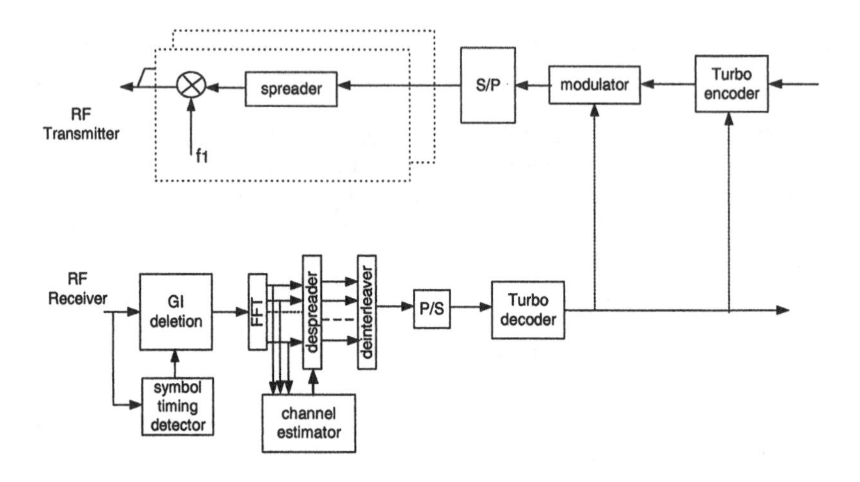

Figure 6.10 A block diagram of NTT DoCoMo's 4G mobile terminal.

In that scenario, typical HSDPA features discussed in Chapter 4, such as link adaptation and HARQ, will also be applied to a new transport channel called OFDM-HS-PDSCH. The current assumption is that network access will be performed through the existing WCDMA carriers, and a handover to the OFDM carrier will be performed when needed. Consequently, the mobile terminal needs to support WCDMA and to have OFDM HS-DSCH receiving capabilities. In the call setup stage, the WCDMA link would be used to achieve the initial network connection. When there is a requirement for higher data rate services, the HS-DSCH mode may be initiated, using either the WCDMA downlink carrier (Release 5 HSDPA) or the separate OFDM downlink carrier. The drawback of this approach is that, while both Release 99 and Release 5 HSDPA traffic can be supported over a common RF carrier, the OFDM HSDPA may need a dedicated carrier.

Owing to the widespread adoption of OFDM-based wireless LANs (WLAN), the interworking of UTRAN and WLAN is becoming important and it will most likely be a basic feature of the fourth generation mobile communications systems (see Chapter 9). Introducing OFDM to UTRAN not only can serve the purpose of increasing the downlink throughput but can also ease the integration of WLAN with UTRAN systems in the physical layer. Even if the physical parameters are different, some modules in the transceivers for future UTRAN and WLAN can be shared. To some extent, this proposal is similar to NTT DoCoMo's fourth generation system in the sense that the uplink will be mainly based on DS-CDMA and the downlink will be mainly based on OFDMA.

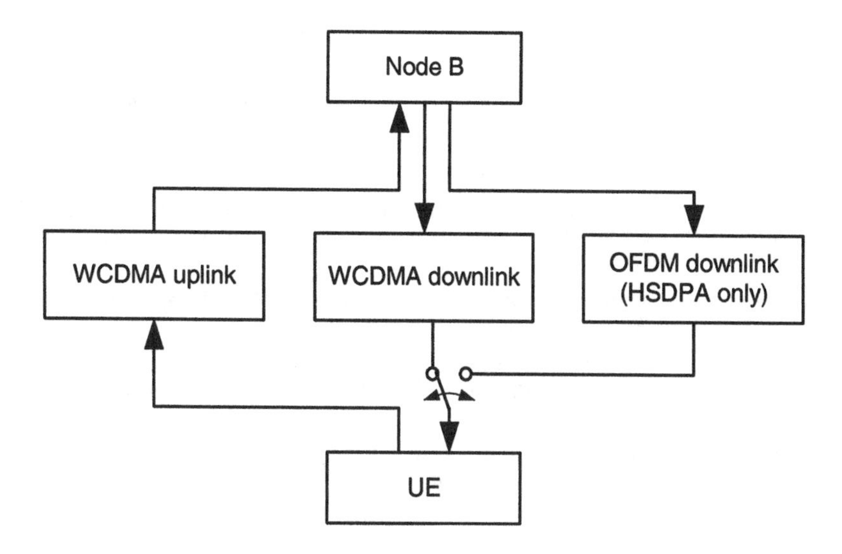

Figure 6.11 Incorporating OFDM/IOTA into UTRAN for HSDPA.

6.6 CONCLUDING REMARKS

As a modulation scheme, OFDM is becoming the primary choice for future wireless communications systems from digital audio broadcasting (DAB), digital video broadcasting (DVB), and wireless LANs (WLAN), to cellular networks. The advantages of OFDMA are that it not only enables a higher data rate, especially in the downlink, but it also offers an attractive path to integrate different kinds of wireless services in the physical layer. The advance from OFDM/QAM to OFDM/IOTA can make the OFDM even more robust.

As a multiple access technology, OFDMA is mature for isolated hot-spot scenarios, but less so for the cellular environment. The current proposals for OFDMA in the cellular environment include frequency hopping spread spectrum OFDM and direct sequence spread spectrum OFDM. The former attempts to avoid intercell interference by employing orthogonal hopping patterns and the latter uses spreading sequences to suppress intercell and intracell interference. The ultimate choice for the new air interface of the fourth generation mobile communications systems will depend not only on technology advances but also on many economical and political factors.

References

[1] L. Hanzo, W. Webb, and T. Keller, *Single and Multi-carrier Quadrature Amplitude Modulation*, New York: John Wiley & Sons, 2001.

[2] http://grouper.ieee.org/groups/802/index.html.

[3] A. E. Jones, T. A. H. Wilkinson, and S. K. Barton, "Coding Scheme for the Reduction of the Peak-to-Mean Envelope Power Ratio of Multicarrier Transmission Schemes," *Electronics Letters*, Vol. 30, No. 25, December 1994, pp. 2098–2099.

[4] C. Tellambura, et al., "Frequency offset correction for HIPERLAN," *IEEE Trans. on Communications*, August 1999, pp. 1137-1139.

[5] H. Bolcskei, "Orthogonal Frequency Division Multiplexing Based on Offset QAM," in *Advances in Gabor Analysis*, H. G. Feichtinger and T. Strohmer, (eds.), Boston, MA: Birkhäuser, 2003, pp. 321–352.

[6] B. L. Floch, M. Alard and C. Berrou, "Coded Orthogonal Frequency Division Multiplexing", *Proc. of the IEEE*, Vol. 83, No. 6, June 1995, pp. 982–996.

[7] M. Kifle, M. Andro, and M. J. Vanderaar, "An OFDM System Using Polyphase Filter and DFT Architecture for Very High Data Rate Applications," NASA/TM—2001–210813.

[8] France Telecom, "IOTA, An Advanced OFDM Modulation for UTRAN Enhancement: Presentation and Potentials," R1-030014, 3GPP TSG RAN-1 Meeting #30, January 2003.

[9] WAVECOM, "Removing the Guard Interval in the OFDM Physical Layer – Introducing the OFDM/OQAM-IOT Physical Layer," R1-03-0087, 3GPP TSG RAN-1 Meeting #30, January 2003.

[10] 3GPP, TR25.892, Feasibility Study for OFDM for UTRAN Enhancement, Ver. 1.1.0, April 2004.

[11] C. Eklund, et al., "IEEE Standard 802.16: A Technical Overview of the WirelessMAN™ Air Interface for Broadband Wireless Access," *IEEE Communications Magazine*, June 2002, pp. 98–107.

[12] http://www.ieee802.org/mbwa/Contributions.html.

[13] http://www.flarion.com.

[14] K. Higuchi, et al., "Overview of Experimental System of Broadband Wireless Access," *IEICE Convention*, 2003.

[15] N. Maeda, et al., "Experiments on Throughput Performance of MCS Sets for VSF-OFDM Broadband Wireless Access Information Link," *IEICE Convention*, 2003.

[16] T. Ohseki, et al., "Proposal of OFDM / MC-CDMA Based Broadband Mobile Communication System," *IEICE Convention*, 2003.

Chapter 7

RAN Architecture Evolution

Currently, the evolution from 2G to 3G mobile communications systems is happening in three planes: radio, network, and services. In the radio plane, the employment of new radio air interfaces enhances the data rates over the air and increases the system capacity. The HSDPA technology for UTRAN and the 1xEV-DO and 1xEV-DV for CDMA2000 further improve the spectrum efficiency for the transmission of data packets. In the network plane, the introduction of the packet-switched (PS) core network makes the transportation of IP-based data traffic much more efficient. In the service plane, there is a fast growing demand on wireless Internet services and other data-centric services such as multimedia and video phones. In particular, the standardization of the IP Multimedia Subsystem (IMS) by 3GPP enables voice over IP (VoIP) services in UTRAN, thus facilitating the convergence of the circuit-switched (CS) and the packet-switched core networks. All these technology advances point to the direction of all-IP networks.

Employing all-IP networks is the most promising way for operators to offer and maintain reliable wireless data services with low cost. All-IP eliminates the need for much of the equipment typically found in legacy wireless networks such as the circuit-switched core networks in GSM and IS-95. This dramatically reduces not only the initial capital expenditures required to deploy the network, but also the expenditures for operating and maintaining it. It is reported by some telecom industry analysts that, using IP technology, the cost for implementing and maintaining a network can be reduced by 50% compared to that associated with a comparable circuit-switched network.

Voice communications are different from data communications. They are delay-and-jitter sensitive but require low bandwidth and can tolerate the occasional losses that are common in wireless communications. On the other hand, data communications are not very sensitive to delay, but they need high bandwidth in a bursty manner and require much higher reliability. This demands that the network allocate high bandwidth to a user, deliver a large burst of information, quickly repair any errors that might occur, and then reallocate the bandwidth to another user. The current UTRAN and CDMA2000 architecture was designed based on a voice-centric paradigm. Therefore, it is relatively inefficient in handling data traffic. Moreover, with this architecture, each RNC is normally connected with dozens or hundreds of node Bs, and a single point of failure at a RNC could lead to a disaster in the network.

To facilitate data communications, the design of the next generation radio access networks (RANs) should be based on the Internet paradigm and the transport network should be built on IP. To increase the network robustness, the RNCs should become leaner, their functionalities should be distributed, and the network topology should evolve from the current star-like one towards a meshed one. To this end, base stations and RNCs should bear a client-server relationship, with the RNC evolving into servers such as user plane servers and radio control servers. Eventually, in addition to their radio functionalities, all the network elements in a RAN will serve as routers and a mobile terminal will be reachable via its IP address. Such a radio access network is referred to as IP-based RAN. In an IP-based RAN, the base station becomes an access router with multiradio air-interface support.

In UTRAN, the GPRS core network that consists of the serving GPRS nodes (SGSNs), the gateway GPRS nodes (GGSNs), and an IP transport network has been introduced to support packet-switched (PS) data traffic. In CDMA2000 1x, the packet core network even offers the support for mobile IP (MIP). The IP-based RAN will extend the IP network from the core network to the radio access network. Eventually, the division between the core network and the radio access network may disappear, and all the functionalities for network control and management will be handled by some standard nodes distributed across the IP network. In such a network, a hierarchical approach to supporting mobility can be employed; the macro mobility is supported by mobile IP, whereas micro mobility is supported by hierarchical mobile IP (HMIP).

In summary, it is expected that UTRAN will evolve into an IP-based RAN architecture. The IP-based RAN needs four major building blocks: (1) IP transport; (2) mobility management, which includes mobile IP (MIP) for macro-mobility management and hierarchical mobile IP (HMIP) for micro-mobility management; (3) distributed network control nodes suited for handling IP traffic, and (4) advanced base stations with multiradio support. In this chapter, the first three issues are discussed and several types of IP-based RAN architecture are presented.

7.1 MOBILE IP

Before starting detailed discussions on mobile IP (MIP), some definitions which are widely used in the Internet Engineering Task Force (IETF) community are given first in this section [1]. A *node* is a device that implements IP and a *router* is a node that forwards IP packets not explicitly addressed to it. A *host* is a node that is not a router. Every node on the Internet has at least one IP address, which is a sequence of 32 bits for IP version 4 and 128 bits for IP version 6. Every IP address has two distinct components, the *subnet prefix* portion and the host portion. A subnet prefix is like a group address and it is a sequence of bits that is identical for all the nodes attached to the same physical link. Each node attached to the link is identified by the host portion in its IP address. The routing mechanism used for the Internet is based on the IP addresses of the connected nodes. By using subnet prefixes, IP addresses define a certain topology relation between different nodes; each subnet includes a number of nodes belonging to either an organization or a geographical area and packets destined for a node are delivered to the subnet first and then to the node itself.

A *mobile node* is a node that can change its point of attachment from one link to another, while still being reachable via its home address. A *home address* is an IP address assigned to a mobile node and it is used as the permanent address of the mobile node. This address is reachable on the mobile node's home link that is defined as the link on which a mobile node's home subnet prefix is defined. Standard IP routing mechanisms deliver packets destined for a mobile node's home address to its home link. A *foreign subnet prefix* is any IP subnet prefix other than the mobile node's home subnet prefix and a foreign link is any link other than the mobile node's home link. A *care-of address* is a routable address associated with a mobile node while visiting a foreign link. Among the multiple care-of addresses that a mobile node may have at any given time, the one registered with the mobile node's home agent is called its "primary" care-of address. A home agent is a router on a mobile node's home link with which the mobile node registers its current care-of address. While the mobile node is away from home, the home agent intercepts packets on the home link destined to the mobile node's home address, encapsulates them, and tunnels them to the mobile node's primary care-of address.

The current Internet is built on the Internet protocol version 4 (IPv4). Despite its great worldwide success, IPv4 has two main drawbacks. First, the IPv4 uses a limited 32-bit address space and, as such, it is estimated by IETF that IPv4 addresses will run out in a few years. Although the address shortage problem can be mitigated to some extent by the network address translation (NAT) technique, NAT hinders network management and compromises the level of network security. Furthermore, NAT is not well suited for the future of ubiquitous mobile communications in which each mobile device will potentially need at least one IP address. Second, In IP version 4, there is no adequate support for security, mobility, and quality of services (QoS). For these reasons, the Internet Protocol version 6 (IPv6) is being developed by the IETF for the next generation Internet, and IPv6 has been adopted by numerous standardization bodies for mobile communications such as 3GPP [2]. IPv6 has a 128-bit address space, and it includes two security features in the core protocol stack: encryption of packets for data privacy and authentication of the sender of packets. Further, IPv6 implements two QoS control techniques, resource reservation and packet prioritization, which make it possible to carry real-time traffic.

The design of the current Internet protocol (IPv4) was based on the assumption that the every node has a fixed point of attachment to the Internet. If a node is moved from one place to another, it is generally necessary to change its address; otherwise, it cannot be reached by any other nodes, and packets destined to the node would get lost. On the other hand, if a node changes its IP address every time it moves to a new link, it would not be able to maintain its existing transport and higher-layer connections. As a solution, mobile IP (MIP) has been developed by the IETF to provide mobility support in the Internet protocol. Since this book is focused on future mobile communications networks, the following discussion will be focused on mobile IPv6 [3].

Mobile IP enables a mobile node to move from one link to another without changing its home address. In a network with mobile IP support, all packets destined to the mobile node can be routed to it using its home address regardless of its current point of attachment to the Internet. The movement of a mobile node away from its home link is thus transparent to the transport and higher-layer protocols and applications. Mobile IP is suitable for mobility support across both homogeneous and heterogeneous networks. For example, mobile IPv6 facilitates node movement from one Ethernet segment to another as well that from an Ethernet segment to a wireless LAN cell, with the mobile node's IP address remaining unchanged. Before considering the operation of mobile IP, it should be noted that mobile IP is mainly aimed at supporting the network-layer mobility management. When applied to cellular networks, some link layer mobility management techniques such as handover are still needed. This topic will be discussed further in Section 7.3.

A mobile node is always expected to be addressable at its home address, being attached to its home link or foreign link. While a mobile node is at home, packets addressed to its home address are routed to the mobile node's home link using conventional Internet routing mechanisms. If a mobile node is attached to some foreign link away from home, it is addressed instead at one or more care-of addresses. The mobile node can acquire its care-of address through standard IPv6 mechanisms. As long as the mobile node stays in this location, packets addressed to this care-of address will be routed to the mobile node.

The association between a mobile node's home address and the care-of address is known as a "binding" for the mobile node. When away from home, a mobile node registers its primary care-of address with a router on its home link, requesting this router to function as the home agent for the mobile node. The mobile node performs this binding registration by sending a *Binding Update* message to the home agent. The home agent replies to the mobile node by returning a *Binding Acknowledgement* message. A node communicating with a mobile node is referred to as a *correspondent node* of the mobile node, and the correspondent node may itself be either a stationary node or a mobile node. There are two possible modes for communications between a mobile node and a correspondent node.

- The first mode, *bidirectional tunneling*, does not require mobile IP support from the correspondent node and is available even if the mobile node has not registered its current binding with the correspondent node. Packets from the correspondent node are routed to the home agent and then tunneled to the mobile node, and this is called triangular routing. Packets to the correspondent node are tunneled from the mobile node to the home agent (reverse tunneled) and then routed normally from the home network to the correspondent node. In this mode, the home agent intercepts all IP packets addressed to the mobile node's home address on the home link. Each intercepted packet is tunneled to the mobile node's primary care-of address by using encapsulation.
- The second mode, *route optimization*, requires the mobile node to register its current binding with the correspondent node. Packets from the correspondent node can be routed directly to the care-of address of the mobile node. When sending a packet to any IPv6 destination, the correspondent node checks its cached bindings for an entry for the packet's destination address. If a cached binding for this destination address is found, the node uses a new type of IPv6 routing header to route the packet to the mobile node by way of the care-of address indicated in this binding.

Routing packets directly to the mobile node's care-of address avoids triangular routing and allows the shortest communications path to be used. It also eliminates congestion at the mobile node's home agent and home link. In addition, the impact of any possible failure of the home agent or networks on the path to or from it is reduced. When routing packets directly to the mobile node, the correspondent node sets the destination address in the IPv6 header to the care-of address of the mobile node. A new type of IPv6 routing header is also added to the packet to carry the desired home address. Similarly, the mobile node sets the source address in the packet's IPv6 header to its current care-of addresses. The mobile node can add a new IPv6 home address destination option to carry its home address. The inclusion of home addresses in these packets makes the care-of address transparent above the network layer.

7.2 FAST HANDOVER IN MOBILE IPv6

Handover refers to the process in which a mobile terminal changes its serving cell. In mobile communications systems, a mobile terminal can be assigned with one or more IP addresses and therefore it can be treated as a mobile node. With each base station serving as a radio *access router* (AR), the base stations become the points of attachment to the Internet. Mobile IP enables a mobile node to maintain connectivity to the Internet during its handover from one access router to another (see Figure 7.1). During the handover process, there is a time period when the mobile node is unable to send or receive IP packets both due to link switching delay and IP protocol operations. This time period is referred to as handover latency. In many instances, the handover latency resulting from standard mobile IPv6 handover procedures can be greater than what is acceptable to support real-time or delay-sensitive traffic. To reduce the handover latency, a fast handover protocol has been introduced to mobile IPv6 [4]. It enables a mobile node to send packets as soon as it detects a new link, and to deliver packets to a mobile node as soon as the presence of the mobile node is detected by the new access router. It allows a mobile node to keep using its existing care-of address until it establishes itself as a mobile IP endpoint on its new access router.

7.2.1 Fast Handover Protocol

The fast handover protocol is aimed at setting up a tunnel between the previous and the new access routers to enable the mobile node to send and receive IP packets while it is establishing itself as a mobile IPv6 endpoint with the new access router. This tunnel establishment can be triggered by the mobile node requesting a handover. It can also be triggered by the previous access router and, in this case, the tunnel establishment is referred to as network-initiated. Once the tunnel is established, packet forwarding on the tunnel to the mobile node begins when the previous access router receives a *Fast Binding Update* message from the mobile node. There are three phases in the protocol operation: handover initiation, tunnel establishment, and packet forwarding.

7.2.1.1 Handover Initiation

The fast handover protocol is initiated when a decision to handover the mobile node to a new access router is made. The trigger for protocol initiation may arrive from the specific link layer or a decision based on preconfigured rules that command a handover. As a response to such a trigger, the mobile node requests its previous access router to assist in handover by sending a *Router Solicitation for a Proxy* (RtSolPr) message, in which it includes the link-layer identifier (such as a base station ID) of its prospective attachment point. Following the message, the previous access router sends a *Proxy Router Advertisement* (PrRtAdv) message, which provides the link-layer address and network prefix information about the new access router. In a handover initiated by the network, the previous router sends a PrRtAdv message without the mobile node sending a RtSolPr message. It provides the parameters necessary for the mobile node to send IP packets, as well as the network prefix for the mobile node to formulate a prospective new care-of address. The purpose of RtSolPr is to request parameters necessary for the mobile node to be able to send packets immediately upon connecting to the new access router. The purpose of PrRtAdv is to supply these parameters and the network prefix information that allows the mobile node to formulate a new care-of address.

7.2.1.2 Tunnel Establishment Between the Access Routers

After receiving a PrRtAdv message, the mobile node sends a *Fast Binding Update* (FBU) to the previous access router. The mobile node may also send an FBU after attaching to the new access router. This FBU message associates the mobile node's previous care-of address with the new access router's IP address so that packets arriving at the previous access router can be tunneled to the new access router. In response, the previous access router sends a *Handover Initiate* (HI) message to the new access router. The HI message serves two purposes. First, it initiates establishing a bidirectional tunnel between the two routers so that the mobile node can continue using the previous care-of address for its existing sessions. Second, it serves to verify if the new care-of address can be used on the new access router's link. Upon processing the HI message, the new access router sets up a route for the mobile node's previous care-of address and responds with a *Handover Acknowledge* (HACK) message. Upon receiving an HACK message, the previous router sends a *Fast Binding Acknowledgment* (FBACK) message to the mobile node. The FBACK message confirms whether the new care-of address can be used, and after which the mobile node can use the new care-of address of the new link.

A bidirectional tunnel is established between the two access routers for the following purpose. Since the mobile node cannot use the new care-of address (NCoA) before it completes its binding update with its home agent and the correspondent nodes, it has to use previous care-of address. To this end, the new access router (NAR) tunnels packets sent from the mobile node with the previous care-of address (PCoA) as the source IP address to the previous access point (PAR). Until the correspondent nodes establish a binding cache entry for the new care-of address, they continue to send packets to the previous care-of address. The PAR tunnels these packets to the NAR, which then forwards them to the mobile node.

7.2.1.3 Packet Forwarding

The mobile node sends a *Fast Binding Update* (FBU) message, preferably prior to disconnecting its link. Since a gratuitous PrRtAdv message indicates a network-controlled handover, the mobile node sends FBU immediately as a response. The mobile node uses PCoA as its source IP address in this message. If the mobile node is unable to send an FBU before leaving the PAR, it will do so as soon as it regains connectivity with the new access router. This allows the PAR to actually start tunneling packets meant for the mobile node's PCoA.

When the PAR receives an FBU, it waits until the requested handover is accepted by the new access router as indicated in the HACK message status code. It verifies that the source IP address in the FBU is the PCoA for which the PAR has forwarded packets previously. Then it creates a tunnel for forwarding packets meant for the PCoA to the new access router. Finally, packets intended for the mobile terminal are forwarded from the previous access router to the new access router.

As soon as the mobile node establishes link connectivity with the new access router, it sends a *Fast Neighbor Advertisement* (FNA) message, which is encoded as an option in the *Router Solicitation* (RS) message. Only after it receives a confirmation to use NCoA in a *Router Advertisement* (RA) with the *Neighbor Advertisement Acknowledge* (NAACK) option, the mobile node starts using the NCoA. If the mobile node has already received confirmation to use NCoA via FBACK, it should include the FNA option anyway to announce its presence to the new access router.

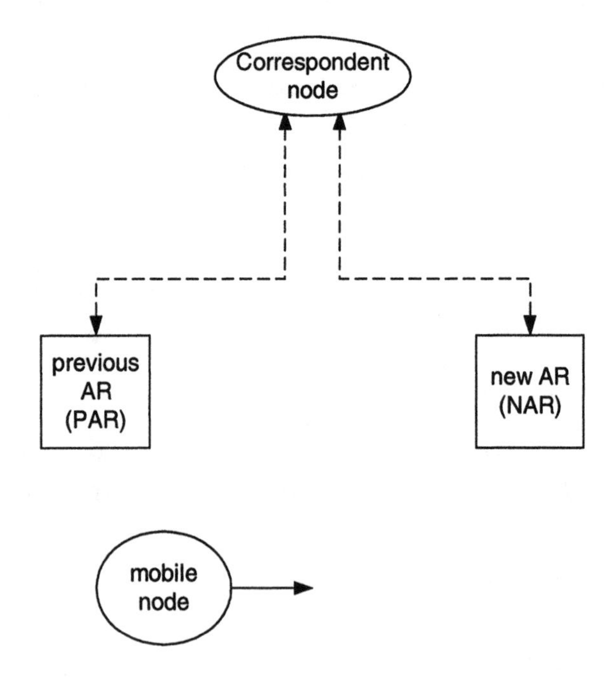

Figure 7.1 Illustration of handover.

Once the mobile node completes the binding update procedure with its correspondent node, it will start using NCoA as its source IP address. In any case, the bidirectional tunnel between the access routers has a lifetime of BT_LIFETIME, after which packets using PCoA as source IP address will be discarded unless there is optional protocol support for tunnel lifetime renewal. The process of fast handover is depicted in Figure 7.2.

7.2.2 Three-Party Handover

The mobile node may move from the new access router to another access router (NAR') before completing its handover and performing binding updates to its correspondents. If the mobile node moves before configuring a NCoA on the new access router, the PAR would still be considered its default router when it is connected to NAR'. Therefore, the mobile node should send an FBU to the PAR to set up a tunnel between the PAR and the NA'. On the other hand, if the mobile node is able to configure a NCoA on the new access router but is unable to update all or a partial list of its correspondents, the mobile node may send an FBU to PAR. This FBU may be in addition to the FBU it sends to the new access router. The result in this case is that the mobile node should be able to receive packets arriving at PCoA and NCoA through potentially different tunnels. If the mobile node returns to its PAR without completing mobile IPv6 updates, it sends an FBU with lifetime set to zero so the PAR can disable any outgoing tunnels.

7.3 HMIPV6

Mobile IP is well suited for nomadic communications in which a terminal is not on the constant move. For mobile communications, however, a further improvement to mobile IP is needed to support the so-called micromobility, which refers to movement between sectors and cells in a radio access network.

Mobile IPv6 allows nodes to move within the Internet topology while maintaining reachability and ongoing connections between the mobile and correspondent nodes. To do this, every time it moves, a mobile node sends binding updates to its home agent and all correspondent nodes with which it communicates. However, authenticating binding updates requires approximately 1.5 round-trip times between the mobile node and each correspondent node. In addition, one round-trip time is needed to update the home agent, which can be done simultaneously while updating correspondent nodes.

In a mobile communications system, there can be frequent handovers between radio access routers for IP-based mobile terminals. If the mobile IPv6 described in the previous sections was used, a binding update would be needed every time a handover to a new access router is performed, and the round-trip delays would introduce frequent disruptions to the communications between the mobile node and the correspondent node. Moreover, the mobile terminal would send and receive a large number of signaling messages over the air interface to all correspondent nodes and the home agent. To improve the performance of mobile IPv6 in such scenarios and to limit the amount of mobile IPv6 signaling outside the local domain, a new local node, *mobility anchor point* (MAP), is introduced to mobile IP, thus resulting in the hierarchical MIPv6 (HMIPv6) [5].

A MAP is essentially a local home agent and it can be located at any level in a hierarchical network of routers including the radio access router, and it aggregates the outbound traffic of mobile nodes. With the introduction of the MAP, the mobile node sends binding updates to the local MAP rather than the correspondent node and home agent, which are typically much further away, thus reducing the signaling traffic in the mobile network. Furthermore, independent of the number of correspondent nodes with which the mobile node is communicating, only one binding update message needs to be transmitted by the mobile node before the traffic from the home agent and all correspondent nodes is rerouted to its new radio access router.

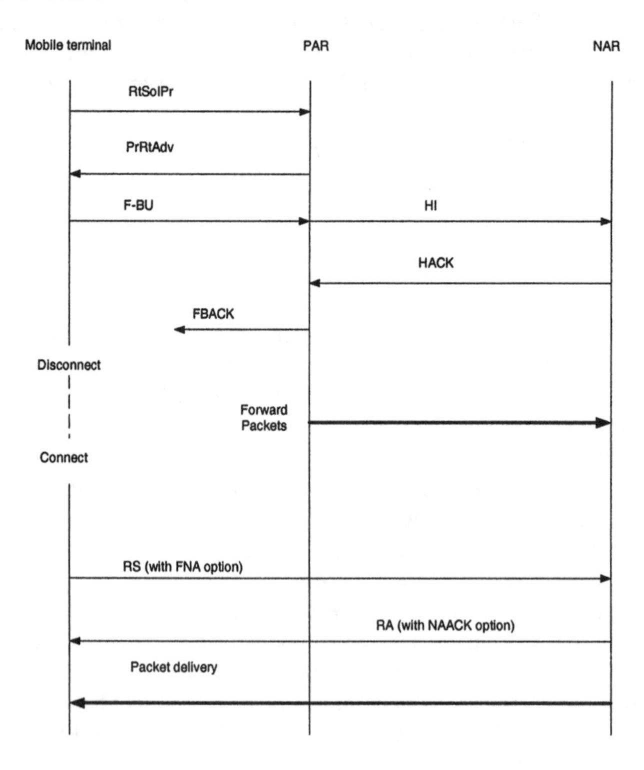

Figure 7.2 Illustration of the fast handover protocol.

A MAP is a router located in a network visited by the mobile node and there can be one or more MAPs. As a consequence of the hierarchical structure, two care-of addresses are introduced into HMIPv6. A regional care-of address (RCoA) is an address on the MAP's subnet and it is obtained by the mobile node from the visited network. It is autoconfigured by the mobile node when receiving the MAP option. A local care-of address (LCoA) is the on-link care-of address configured on the mobile node's interface based on the subnet prefix advertised by its default radio access router. A *local binding update* is the update sent by the mobile node to the MAP in order to establish a binding between the RCoA and the LCoA.

A mobile node entering a MAP domain will receive *router advertisements* containing information on one or more local MAPs. The mobile node can bind its current on-link CoA with a RCoA address on the MAP's subnet. Acting as a local home agent, the MAP will receive all packets on behalf of the mobile node it is serving and will encapsulate and forward them directly to the mobile node's current address. If the mobile node changes its current address within a local MAP domain (LCoA), it only needs to register the new address with the MAP. Hence, only the regional CoA (RCoA) needs to be registered with correspondent nodes and the home agent. The RCoA does not change as long as the mobile node moves within a MAP domain. A MAP domain's boundaries are defined by the access routers advertising the MAP information to the attached mobile nodes.

It should be noted that the HMIPv6 concept is simply an extension to the mobile IPv6 protocol. An HMIPv6-aware mobile node with an implementation of Mobile IPv6 only chooses to use the MAP when discovering such capability in a visited network. However, in some cases the mobile node may prefer to simply use the standard mobile IPv6 implementation. For instance, the mobile node may be located in a visited network within its home site. In this case, the home agent is located near the visited network and could be used instead of a MAP. In this scenario, the mobile node would only update the home agent whenever it moves.

The network architecture shown in Figure 7.3 gives an example of the use of the MAP in a visited network. Upon arrival in a visited network, the mobile node will discover the global address of the MAP. This address is stored in the access routers and communicated to the mobile node via *router advertisements* (RAs). This is needed to inform mobile nodes about the presence of the MAP. In the discovery phase, the mobile node is also informed of the distance of the MAP from the mobile node. For example, the MAP function could be implemented as shown in Figure 7.3 and at the same time also in AR1 and AR2. In this case the mobile node can choose the first hop MAP or one further in the hierarchy of routers. The process of MAP discovery continues as the mobile node moves from one subnet to the next. As the mobile node roams within a MAP domain, it receives an announcement from access routers about the same MAP address or addresses. If a change in the advertised MAP's address is received, the mobile node acts on the change by performing movement detection and sending the necessary binding updates to its home agent and the correspondent nodes.

If the mobile node is not HMIPv6-aware, then no MAP discovery will be performed, thus resulting in the mobile node using the mobile IPv6 protocol for its mobility management. On the other hand, if the mobile node is HMIPv6-aware, the mobile node will first need to register with a MAP by sending it a binding update containing its home address and on-link care-of address (LCoA). The MAP stores this information in its binding cache to be able to forward packets to their final destination when received from the different correspondent nodes or home agents.

The mobile node will always need to know the original sender of any received packets to determine if route optimization is required. This information will be available to the mobile node since the MAP does not modify the contents of the original packet. Normal processing of the received packets will give the mobile node the necessary information.

To use the network bandwidth in a more efficient manner, a mobile node may decide to register with more than one MAP simultaneously and use each MAP address for a specific group of correspondent nodes. For example, in Figure 7.3, if the correspondent node happens to exist on the same link as the mobile node, it would be more efficient to use the first hop MAP (in this case, assume it is AR1) for communications between them. This will avoid sending all packets via the highest MAP in the hierarchy and hence result in a more efficient usage of network bandwidth.

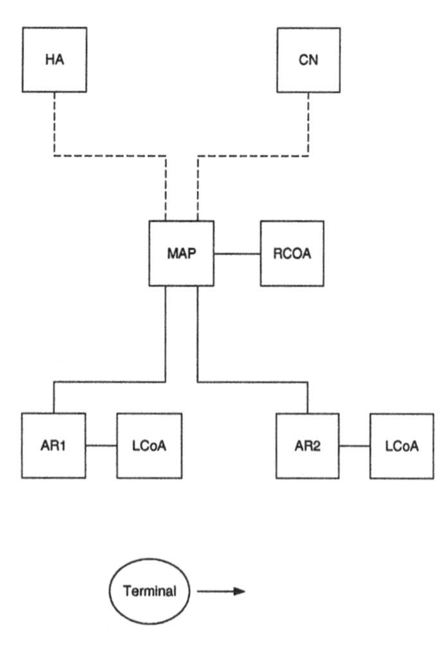

Figure 7.3 Hierarchical mobile IPv6 domain.

7.3.1 Mobile Node Operation

In HMIPv6, the mobile node has two addresses, an RCoA on the MAP's subnet and an on-link CoA (LCoA). This RCoA is formed by combining the mobile node's interface identifier and the subnet prefix received in the MAP option. This protocol requires special treatment by the mobile nodes only. The home agent is unchanged. The MAP performs the function of a local home agent that binds the mobile node's RCoA to an LCoA.

When a mobile node moves into a new MAP domain, it needs to configure two CoAs: an RCoA on the MAP's subnet and an on-link CoA (LCoA). After forming the RCoA based on the prefix received in the MAP option, the mobile node sends a local binding update to the MAP. The local binding update is one that specifies the mobile node's RCoA in the home address option. The LCoA is used as the source address of the binding update. This binding update will bind the mobile node's RCoA to its LCoA. Upon registering with the MAP, the mobile node will register its new RCoA with its home agent by sending a binding update that specifies the binding between the RCoA and the home address. The mobile node also sends a similar binding update to its current correspondent nodes.

In order to accelerate the local home agent handover between MAPs, a mobile node may send a local binding update to its previous MAP specifying its new LCoA. Packets in transit that reach the previous MAP are then forwarded to the new LCoA. The MAP will receive packets addressed to the mobile node's RCoA (from the home agent or correspondent nodes). Packets will be tunneled from the MAP to the mobile node's LCoA. The mobile node will decapsulate the packets and process them in the normal manner.

When the mobile node moves locally (i.e., its MAP does not change), it only registers its new LCoA with its MAP. In this case, the RCoA stays unchanged. It should be noted that a mobile node may send a binding update containing its LCoA to correspondent nodes which are connected to its same link. Packets will then be routed directly without going through the MAP.

7.3.2 MAP Operations

The MAP acts like a home agent; it intercepts all packets addressed to all the registered mobile nodes and tunnels them to the corresponding LCoAs. A MAP has no knowledge of the mobile node's home address. The mobile node sends a local binding update to the MAP to inform the MAP that the mobile node has formed an RCoA by using the RCoA as the home address in the binding update. If successful, the MAP must return a binding acknowledgment to the mobile node, indicating a successful registration. Otherwise, the MAP must return a binding acknowledgment with the appropriate fault code. The MAP is able to accept packets tunneled from the mobile node, with the mobile node being the tunnel entry point and the MAP being the tunnel exit point. The MAP then acts as a home agent for the RCoA. Packets addressed to the RCoA are intercepted by the MAP, encapsulated, and routed to the mobile node's LCoA. It should be pointed out that the support of HMIPv6 is completely transparent to the home agent's operation. Packets addressed to a mobile node's home address are forwarded by the home agent to its RCoA.

HMIPv6 provides a flexible mechanism for local mobility management within a visited network. A MAP can exist on any level in a hierarchy, including the access router. Several MAPs can be located within a hierarchy independently of each other. In addition, overlapping MAP domains are also allowed and recommended. Both static and dynamic hierarchies are supported. In [5], it is recommended that the mobile node in a hierarchical mobile IP network should be "eager" to perform new bindings and "lazy" in releasing existing bindings. This means that the mobile node should register with any "new" MAP advertised by the AR (eager). The mobile node should not release existing bindings until it no longer receives the MAP option or the lifetime of its existing binding expires (lazy). This eager-lazy approach is useful in providing a fallback mechanism in case of the failure of one of the MAP routers, as it would reduce the time it takes for a mobile node to inform its correspondent nodes and home agent about its new CoA. It should be pointed out that allowing more than one MAP within a network should not imply that the mobile node forces packets to be routed down the hierarchy of MAPs. Placing more than one MAP above the AR can be used for redundancy and as an optimization for the different mobility scenarios experienced by mobile nodes. The MAPs are used independently from each other by the mobile node.

In terms of the distance based selection in a network with several MAPs, a mobile node may choose to register with the furthest MAP to avoid frequent re-registrations. This is particularly important for fast mobile nodes that will perform frequent handoffs. In this scenario, the choice of a further MAP would reduce the probability of having to change a MAP and informing all correspondent nodes and the HA.

7.4 IP TRANSPORT IN UTRAN

3GPP has taken the first step in the UTRAN architecture evolution path to introduce the IP transport network [6]. The IP transport network in UTRAN is comprised of a set of routers and links connecting network elements implementing UTRAN functions including node Bs, RNCs, and network management platforms. The network is responsible for transporting user plane, control plane, and O&M data between the network elements.

In an IP transport network, a host such as a node B or a RNC sees only those routers that are directly accessible, which are known as edge routers. In most cases, there is only one such router available for a network element. Due to the location of node Bs, the physical medium between a node B and the edge router is often bandwidth limited. Figure 7.4 shows an example of an IP transport network.

7.5 IP IN UMTS CORE NETWORKS

7.5.1 UTRA Core Network

Both the second and third generation mobile communications systems consist of radio access networks and core networks. In the UTRAN network, the generalized packet radio service (GPRS) core network is added to the legacy GSM core network, which was designed to handle circuit-switched traffic only, to deal with data traffic. The GPRS core network is based on IP and it consists of two types of network elements, the serving GPRS support node (SGSN) and the gateway GPRS support node (GGSN). The SGSN is the node serving mobile terminals, and its main functionalities include mobility management, traffic routing, authorization, and authentication. In particular, the following tasks are required: detection of new mobile terminals in the serving area; mobile user authentication, authorization and admission control; sending and receiving data packets to and from mobile terminals; and recording of the location of mobile terminals in its serving area. The GGSN is the gateway of an operator's IP network to external networks such as the World Wide Web (WWW) and virtual private networks (VPN). It is responsible for address mapping and GPRS tunneling. The SGSN and GGSN form an operator's IP backbone. The GPRS protocol for the user plane is shown in Figure 7.5 and an important layer to notice is the GPRS tunneling protocol, GTP. GTP is a tunneling protocol running over UDP/IP and it is used to route packets between RNC, SGSN, and GGSN within the same or between different UTRAN backbone(s). A GTP tunnel is identified at each end by a tunnel endpoint identifier (TEID).

It should be noted that, in the current UTRAN architecture, the GTP is terminated at the RNC. The radio access network is operated by the radio interface protocols. IP packets to and from the terminal are tunneled through the network and they are not routed directly at the IP level. For every mobile terminal, one GTP_C tunnel is established for signaling and a number of GTP_U tunnels are established for user traffic. In fact, mobile terminals for UTRAN do not have fixed IP addresses. To connect a laptop computer UTRAN, a handset or modem is needed to perform the protocol conversion as shown in Figure 7.6, where PPP stands for point-to-point protocol [1].

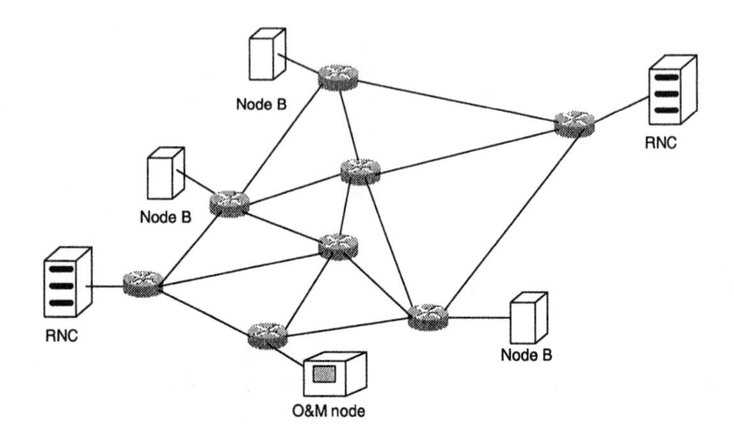

Figure 7.4 IP transport in UTRAN.

In UTRAN Release 6 architecture, the IP multimedia subsystem (IMS) will be introduced (see Figure 7.7). IMS core network and mobile terminals will exclusively support IPv6 for connection to services. IMS makes it possible to support multimedia services and voice over IP, thus integrating the packet-switched core network (PS-CN) and the circuit-switched core network (CS-CN).

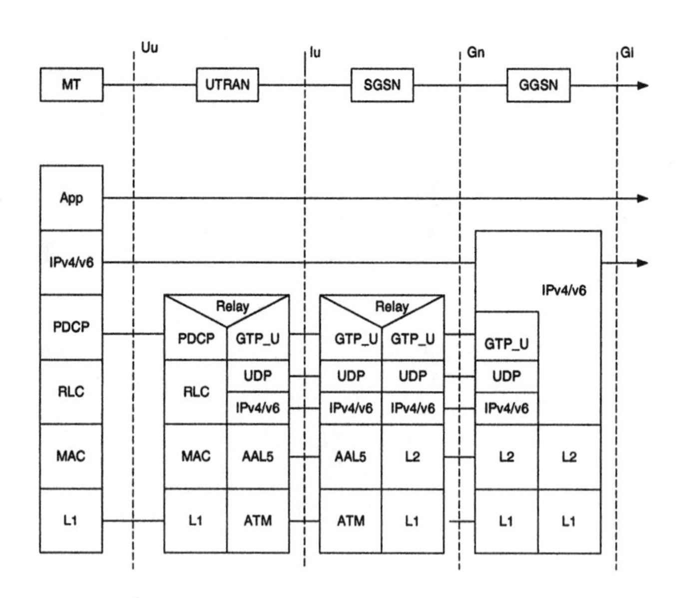

Figure 7.5 GPRS user plane protocol.

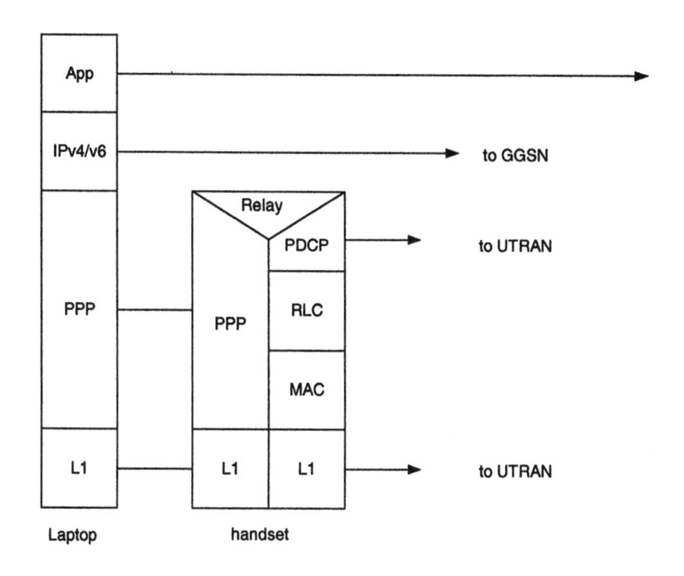

Figure 7.6 Connecting a laptop computer to GPRS.

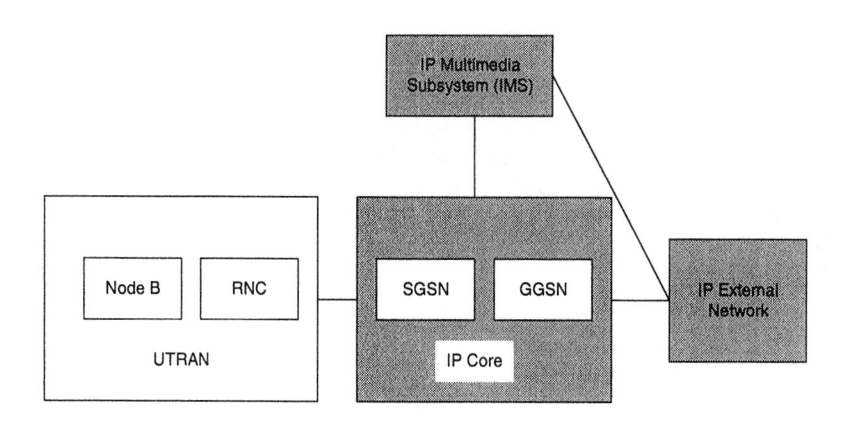

Figure 7.7 UTRAN architecture for Release 6.

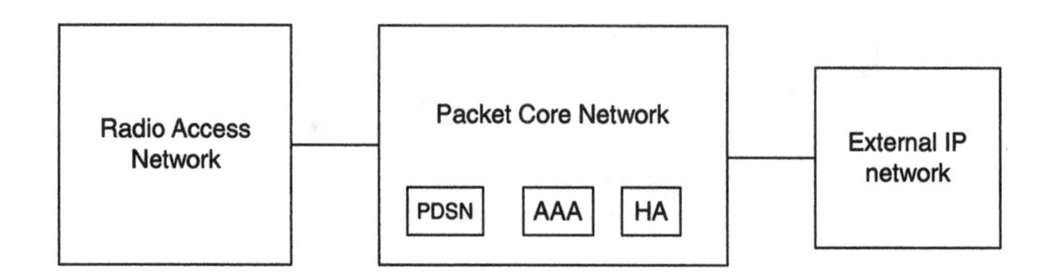

Figure 7.8 CDMA2000 core network.

7.5.2 CDMA2000 1x Core Network

To enable packet handling using IP technology, a packet core network (PCN) has been introduced into CDMA2000 1x that supports mobile IPv4. The CDMA2000 1x packet core network consists of three elements: the packet data serving node (PDSN), the home agent (HA), and the authentication, authorization, and accounting (AAA) server (see Figure 7.8). The packet data serving node (PDSN) functions as a connection point for the radio and IP/ATM networks. Point-to-point Protocol (PPP) links are established, maintained, and terminated here. In addition, the PDSN delivers foreign agent (FA) functionality to register and facilitate services for network visitors (roamers). The home agent (HA), in conjunction with the PDSN, authenticates mobile IP registrations from the mobile client and maintains current location information. The HA also performs packet tunneling, a function that receives packets destined for a mobile's permanent address and routes them to the mobile's new temporary address. The AAA server authenticates and authorizes the mobile client, provides user profile and quality of service (QoS) information to the PDSN, and stores accounting data.

It should be noted that, in contrast to the UTRAN core network, the AAA, HA, and FA in the CDMA2000 1x core network are standard IP network functions.

7.6 IP-BASED RAN

Given the current pace of initial UMTS rollout by mobile operators in Europe and the Far East, it is not expected that a dramatic network architecture change will take place in the near future. However, there have been active research and standardization activities in the UTRAN architecture evolution. In the short term, there is the necessity of addressing the issue of RNC product evolution by major mobile network vendors, as the expected surge in data traffic will put great pressure on the first generation RNCs once UTRAN services are adopted by the mass market. As such, infrastructure vendors are facing the choice of various migration paths. Critical issues to be addressed include the introduction of IP-based transport, cost reduction, user plane performance, and footprint reduction. In the long run, it is the general view that future mobile telecommunications networks will evolve into the direction of IP-based RAN as defined in the beginning of this chapter [7]. The motivations to evolve into an IP-based RAN architecture are the following:

- To support IP-based wireless packet data and voice services;
- To reduce the network equipment expenditure and operational cost;
- To increase bearer bandwidth efficiency;
- To increase robustness and scalability;
- To integrate multiple current and future air interfaces.

Under the framework of the UTRAN architecture evolution and wireless Internet, a number of international forums have been working on the IP-based RAN architecture and various proposals have been made [8-13]. Generally speaking, all existing system proposals are aimed at moving away from the centralized architecture and creating a distributed one, so all network elements will effectively become either a host or a router in an IP network. The term "distributed" has two meanings in this context. First, the radio network controllers will be functionally and physically separated into thinner nodes. Second, various network nodes with different functionalities will be geographically distributed to form a logically meshed network. In particular, the new architecture will introduce client-server relationships between the network elements to enable service across many logical nodes and enable the separation of radio control and bearer control functionalities and the separation of call control and mobility management functionalities. Also, cell-dependent radio and bearer control functions will be pushed towards the radio access nodes. As a result, the single point of failure existing in a logical star network is removed and the network elements become slimmer and more economical to manufacture and maintain.

7.6.1 Architecture Changes

Although the current node B architecture will be affected by the UTRAN evolution, it is the RNC which will be affected the most. It is expected that, on the evolution path to IP-based RAN, the following changes will be made in the RNC [10].

Separation of Control from User Traffic

The separation of control from bearer (user) traffic has been an approach used in the public-switched telephone networks. The main reason is that control and bearer packets have different QoS requirements and therefore can be routed with different QoS parameters. Also, volumes of the control and bearer traffic normally grow at different paces. Networks will become more scalable if resources for control and bearer packets can be scaled independently. For this reason, it is expected that, in future radio access networks, the functional entities for handling control and user traffic will be physically separated into different network nodes.

Separation of Call Processing from Mobility Management

Separation of call processing (CP) from mobility management (MM) functionality optimizes the usage of resources. The term mobility here implies the mobility within a radio access network (micro mobility). For example, the MM functionality is only applicable to mobile users, and radio systems may adopt different mobility management strategies, while the call processing (CP) functionality may be shared by a number of air interfaces and even systems. As the first step, it is expected that a common radio resource management (CRRM) node or module will be produced to deal with the mobility of UTRAN, GSM, and wireless LAN (WLAN) users in Europe and the far East.

Client-Server Architecture

The client-server model, which is widely used in corporate computing and Internet information system access, enables a "flat" architecture. Compared with the hierarchical relationship that is currently used, introducing a client-server relationship between radio network controller and the access nodes improves the RAN architecture in both flexibility and scalability. The RNCs will evolve into two servers, typically radio control servers (RCS) and user plane servers (UPS), and the access nodes become their clients. The radio control servers can be a pooled resource for access nodes that span a wide geographical area and employ different air interfaces.

The two key network elements of an IP-based RAN are the user plane server (UPS) and the radio control server (RCS). The user plane server transforms the user data units into radio frames to be transmitted to the base station and vice versa. The radio control server controls the channel access and manages radio resources. The RCS can be further divided into radio access servers and radio resource management servers and, if required, the user plane server can even be pushed into the access node.

Distributed Servers

To further improve flexibility and scalability, and to reduce the operational cost when making network changes, one can distribute thin UPS and RCS servers across the network. System flexibility is improved because the resources are distributed and not dependent on the uptime of a central computing resource. Scalability is improved because the management and processing functions can be enhanced by adding more appropriate servers as the number of access nodes increases.

The distributed architecture can reduce or eliminate the reparenting process in today's RANs. In today's master-slave model, a substantial network configuration effort is required to add an access node to a RNC. Also, moving an access node from one RNC to another requires technicians to decommission the access node and reprovision the access node on another RNC. The process is not very automated. Reparenting is very useful for adjusting networks to accommodate changes in network load. In a distributed architecture the access nodes contain the site-specific information, but little reprovisioning is needed to change the assignment on access nodes from one radio control server to another. The distributed architecture enables automatic provisioning, and the concept of plug-and-play access nodes becomes feasible. The human effort required to manage the distributed networks can be much less than that required for today's hierarchical networks.

Interworking of RAN with the Core Network

To bridge the IP-based radio access network with the core network, one can place a gateway between the two to serve as a mobility anchor point (MAP) as discussed in Section 7.3 and an interworking unit. Alternatively, one can even merge the radio access network with the core network via a common IP transport network, which is a possible approach to harmonize WCDMA and CDMA2000 networks [14]. Since these important issues need to be standardized, all these options are open. Nevertheless, Figure 7.9 shows an example architecture for IP-based RAN.

7.6.2 Potential Benefits of the IP-Based RAN

The potential benefits of an IP-based RAN are numerous [15]. The most important one to mobile operators is the reduction of network operational and maintenance costs. Another benefit, as discussed earlier, is the flexibility, scalability, and ease of upgrade. Further, it allows efficient use of transport resources as switches and routers are shared across the entire radio access network. It reduces the complexity of network management, so the operation and maintenance of the network can be done in an autonomous manner (see Chapter 8). By pushing time critical radio functions closer to the air interface, it enables optimized radio performance. Last but not the least, it allows cost-efficient service definition, provisioning, monitoring, and QoS reporting.

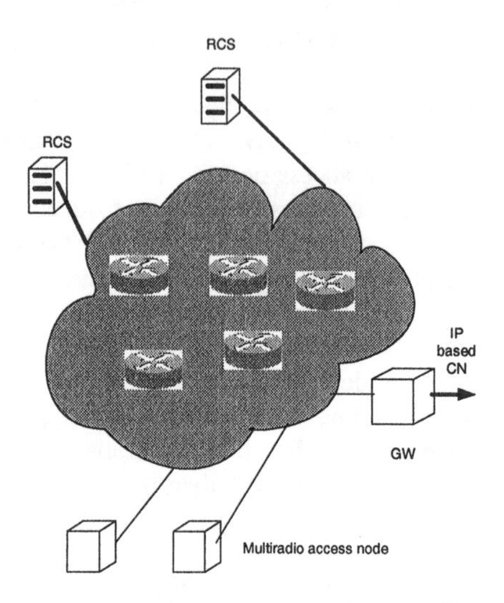

Figure 7.9 An example of IP-based RAN architecture.

The IP architecture model has proven to be very effective for connecting distributed processing nodes. This model has become the dominant architecture for the global Internet as well as intranets. One of the reasons for this broad acceptance of the Internet architecture has been the capability to connect distributed processing nodes via a wide range of physical and link layers. The IP transport layer can be supported by a variety of lower-layer protocols and hence very adaptable to a variety of transport media. The IP transport protocols work well over wide area networks, local area networks, time division multiplexed circuits, synchronous fiber networks, wireless point-to-point links, and asynchronous transfer mode (ATM) circuits. The transport network designer has many degrees of freedom to choose the transport media and lower-level protocols based on criteria such as cost, availability, and bandwidth.

Another reason for the broad deployment of the IP architecture model is that IP network architectures are based on routing rather than switching. In routed networks, the network topology information is autonomously "learned" via routing protocols and distributed throughout the network transport nodes. There is no path state information contained in the transport nodes for any particular session. This allows transport nodes to be added and deleted without affecting other transport or end nodes in the network. This allows simpler fault management and "self-healing" capabilities. Furthermore, path connection/release is not required prior to session initiation or after session termination which facilitates a low overhead session establishment. With the advent of IP-based network capabilities such as dynamic host configuration protocol (DHCP), networks of IP-based clients and servers can be largely self-configured. Distributed processing nodes can be connected or disconnected at any time and the network automatically provides transport-related configuration information. RAN elements on an IP network can be configured for remote management and monitoring using protocols such as simple network management protocol (SNMP) (see Chapter 8).

Remote monitoring and management via open protocols and platforms are a key factor for keeping the network management cost low. The low cost for implementing and maintaining IP networks is also a significant advantage. The widespread use of IP networks can lower both development and operational costs. The high-volume market for IP-based equipment assures a continued cost reduction due to economies of scale. Because of IETF and industry activities, the capabilities of IP-based solutions continue to increase and the cost of the implementations continues to decline. The RAN designer and operator may expect to realize a significant cost and time-to-market benefit by using IP-based networks.

7.7 INDUSTRIAL PROPOSALS

Although the grand view of IP-based RAN is shared by the mobile communications community, the discussion on the detailed evolution path is still in its initial phase [11, 13]. One of the main issues is how far the user plane server should be pushed towards the access nodes (node B). For instance, one can have centralized UPS to perform part of the current RNC functionality; one can have a number of thin and lean UPSs distributed among the network; or one can integrate mini UPSs into the access node. Another issue is the functionality of the gateway between the IP-based RAN and the IP-based core network, or indeed if one should merge the IP-based RAN with the core network. To some extent, this is a matter of choosing the pace of evolution, but such a choice must reflect the market reality in terms of equipment reusability and investment return for mobile operators.

Currently, there are a number of proposals for the evolution of radio access networks [13]. The commonality of the proposed architectures is that the functions of the RNC are decomposed and mapped onto two new types of network entities, radio control servers (RCS) and user plane servers (UPS). The radio control servers mainly perform user-related control functions and coordinate radio resource management. The user plane servers perform cell-related control functions and process radio frames including macro-diversity combining. They also forward user-related control messages between mobile terminals and RCSs. The user plane servers may be moved close to the node Bs, whereas the radio control servers can either be distributed in the network or grouped into clusters.

In the Siemens proposal [11, 13], the Iub interface is kept unchanged from the Release 99 and Release 5 architectures, and it is terminated at the user plane servers. A new interface between the user plane server and the radio control server, Iui, is introduced. Consequently, the Iur and Iu interface are split into Iur-u and Iur-c and Iu-u and Iu-c, respectively. The Iur-u and Iu-u terminate at user plane servers, and Iur-c and Iu-c terminate at the radio control servers, respectively. The user plane servers are controlled by the radio control servers over Iui.

The user plane servers may be moved relatively close to the node Bs while control plane functionality can remain more centralized on radio control servers that might be organized in pools or clusters. The only new type of interface to be defined is the Iui interface between radio control servers and user plane servers. All other interfaces can be derived from existing UTRAN interfaces, which also minimizes the impact of the distributed UTRAN architecture on the external interfaces of the RAN.

Since the Uu interface is not affected in the Siemens proposal, mobile terminals compliant to the current UTRAN standards are fully supported. In fact, even the protocols on the new Iui interface can be derived from existing 3GPP protocols. Mobility mechanisms in the control plane and in the user plane are designed independently from each other. This allows minimizing the number of relocations in the control plane and optimizing traffic flows in the user plane according to the mobility of the terminals.

As a further step in UTRAN architecture evolution, the Iu, Iur, and Iui interfaces can be based on IP transport (see Figure 7.10). In Figure 7.10, the UPSs and RCSs are assumed to have IP routing functionalities so no separate routers are shown. To protect operators' investment, the Iub interface, however, will probably still implement the two transport options, IP and ATM, in order to allow UPSs to be connected to existing node Bs over already deployed ATM-based transport networks. A global picture for future mobile communications systems, which is an integration of different generation of systems, can be found at NEC's Web site [12].

Although based on a different airinterface OFDM, as discussed in Chapter 6, the FLASH-OFDM architecture developed by the U.S. firm Flarion is another interesting proposal for systems beyond 3G (see Figure 7.11) [16]. The FLASH-OFDM system consists of radio routers and an IP packet data network. The radio router is a combination of a base station and an IP access router. The mobility management solution in FLASH-OFDM is based on mobile IP and it supports handover and roaming. One salient advantage of the FLASH-OFDM architecture is its simplicity, since the only wireless-aware element is the radio access router. All other devices in the FLASH-OFDM architecture are standard IP devices working within the standard Internet architecture, with no specialized wireless-specific knowledge, equipment, or mechanisms. Undoubtedly, this type of architecture is well suited for handling packet data.

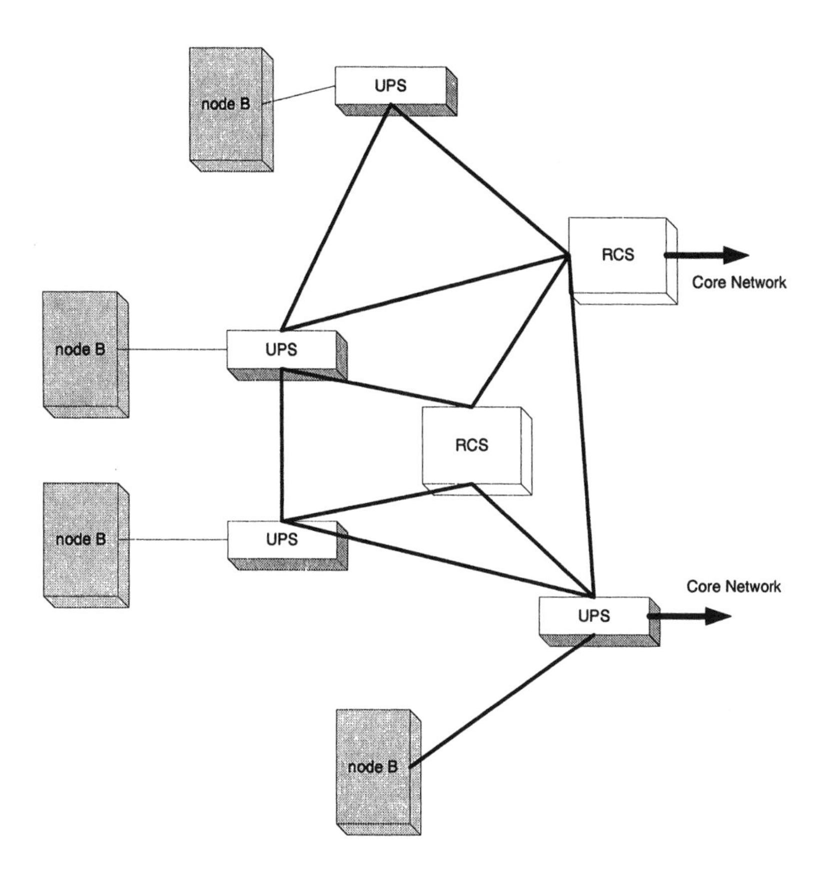

Figure 7.10 A possible architecture for UTRAN evolution.

It should be pointed out that the architecture developed by Flarion is more or less in line with the one being developed by the Japanese forum IP^2 for fourth generation mobile communications systems [14]. Compared with third generation systems, there are several salient features in the IP^2 architecture. First, the core network will be merged with the radio access network. Therefore, radio resource management and session control and management will be dealt with by a unified network. Second, in IP^2, the base stations become radio routers supporting multiple air interfaces, which include WCDMA, wireless LAN (WLAN), wireless personal area networks (WPAN), and possibly a new interface to enable much higher data rates over the air.

7.8 SOFTWARE-DEFINED NETWORK NODE (SDNN)

Currently, although all the third generation mobile communications systems are developed according to 3GPP and 3GPP2 specifications, the designs of the network elements differ from vendor to vendor. To cope with the demand of heavy signal processing and also to reuse vendors' expertise, many components in the base station, radio network controller, and core network are built on application-specific integrated circuits (ASIC). This has served the purpose of early network development and deployment, but the cost of the network elements has been high, and the upgrade and the maintenance can be expensive.

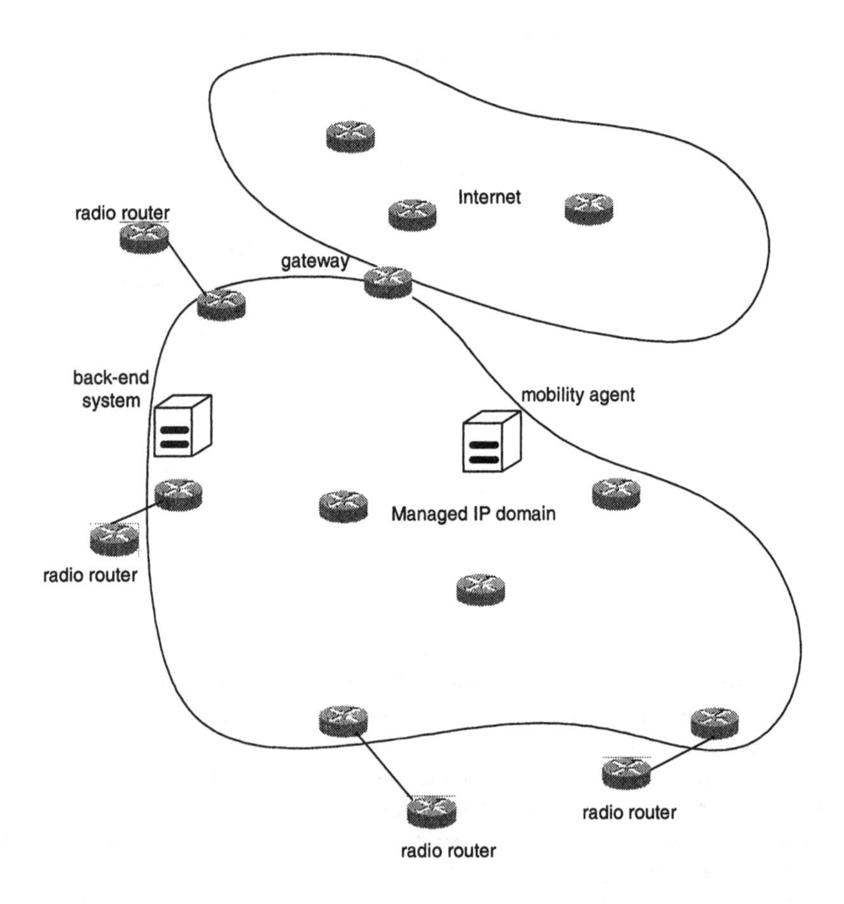

Figure 7.11 Illustration of FLASH-OFDM architecture.

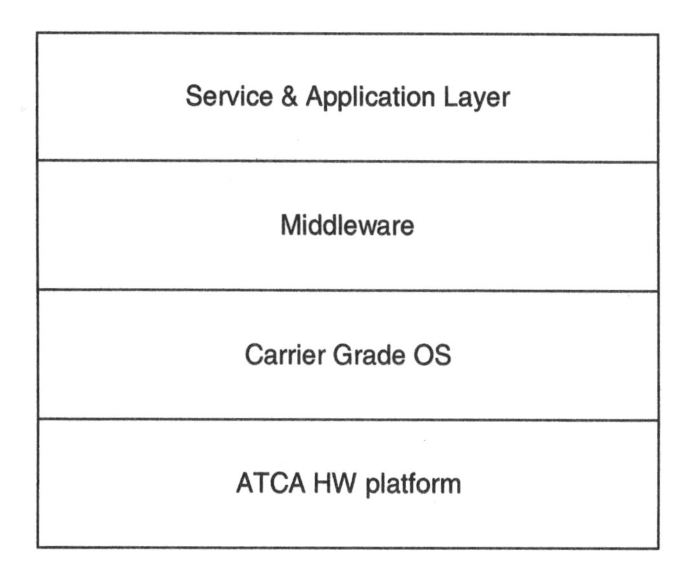

Figure 7.12 Illustration of the generic AdvancedTCA architecture.

Faced with financial pressure on both network vendors and mobile communications operators and technology convergence of communications and computing, the telecommunications and computer industry is moving towards a single modular network platform whose aim is to build standard network modules on the same hardware platform and same operating system, possibly even with the same middleware support [17]. This will result in the so-called software-defined network nodes (SDNO), in which the differences between network elements and the expertise of vendors lie mainly in the software. A great advantage of the software-defined network nodes is the economy of scale, as the same platform can be used by all network vendors in different systems. Currently, an industrial standard in this regard that is referred to as the AdvancedTCA has been ratified by the PCI Industrial Computer Manufacturers Group (PIMC) [18, 19]. The AdvancedTCA specifies the platform architecture for carrier grade telecommunications applications. It is optimized around connectivity requirements of signaling and media gateways. It offers a high level of modularity and reconfigurability. The base specification PICMG 3.0 defines the mechanical form factor, power, and cooling parameters, backplane interconnects, and the system management architecture necessary to construct a compliant backplane, chassis, and plug-in cards. It also defines base fabric protocols for system control management. Subsidiary specifications cover fabric protocols for control and data plane communications. These include PICMG 3.1 for 10/100/1000 Ethernet, PICMG 3.2 for InfiniBand, and PICMG 3.3 for StarFabric technologies. Other fabric definitions are expected to follow in the near future.

The high-level architecture of the software-defined network nodes (SDNO) based on the AdvancedTCA platform is shown in Figure 7.12. It is reported that, using the AdvancedTCA platform, some network vendors such as Lucent Technology are working with Intel to actually design RNCs for UTRAN [19]. Naturally, the platform can also be used for future network nodes such as UPSs and RCSs. Regardless of the details of the RAN evolution path, it is expected that the AdvancedTCA technology will serve as an enabler for the IP-based RAN.

7.9 CONCLUDING REMARKS

The architecture evolution of the radio access networks is being studied by various regional forums and industrial initiatives. To leverage the competitive advantages of different interest groups, different proposals have been made recently. Despite their differences, the grand vision of the future radio access network architecture is pointing in the same direction, a common IP network with distributed nodes and multiradio support, which can be broadly referred to as IP-based RAN. In the IP-based RAN, mobile IP (MIP) and hierarchical mobile IP (HMIP) are expected to play important roles in macro- and micro-mobility management. It may be true that the vision of IP-based RAN as we have defined it will not materialize until the fourth generation systems, but, given the widespread deployment of GSM, WCDMA, and CDMA2000 systems, some pragmatic evolution paths are needed. It is expected that the architecture evolution will eventually offer the following economic advantages:

- Significant savings in infrastructure and terminal cost and operational expenditure;
- Catering for a variety of air interfaces and systems;
- Ease of network operation and maintenance;
- Leveraging the Internet's full economies of scale in infrastructure, devices, applications, and content;
- Greater spectrum efficiency;
- Satisfactory quality of services (QoS) to meet the requirement of a variety of applications and users.

References

[1] J. D. Solomon, *Mobile IP: The Internet Unplugged*, Upper Saddle River, NJ: Prentice Hall, 1998.

[2] M. Wasserman, "Recommendation for IPv6 in Third Generation Partnership Project (3GPP) Standards," *RFC3314*, IETF, September 2002.

[3] D. Johnson, C. Perkins, and J. Arkko, "Mobility Support in IPv6," Internet-Draft, February 26, 2003.

[4] R. Koodli, "Fast Handovers for Mobile IPv6," Internet-Draft, draft-ietf-mobileip-fast-mipv6-06.txt, March 2003.

[5] H. Soliman, et al., "Hierarchical Mobile IPv6 Mobility Management (HMIPv6)," Internet-Draft, October 2002.

[6] 3GPP TR 25.933, V5.2, IP Transport in UTRAN.

[7] W. Mohr and W. Konhauser, "Access Network Evolution Beyond Third Generation Mobile Communications," IEEE *Communications Magazine*, December 2000, pp. 122-133.

[8] M. Schopp, "Towards an IP-based RAN, A Cornerstone for UTRAN Evolution," 3GPP RAN Workshop on UTRAN Evolution, Helsinki, February 2000.

[9] A. Colley, "Evolving the Radio Access Network," *All IP Mobile Network Conference*, London England, May 2001.

[10] Mobile Wireless Internet Forum (MWIF), "OpenRAN Architecture in 3rd Generation Mobile Systems," Technical Report MTR-007, Release v0.3.0, March 2001.

[11] Siemens AG, "Proposed Architecture for UTRAN Evolution," 3GPP TSG-RAN WG3 Meeting #36, TDOC R3-030678.

[12] http://www.nec.co.jp/3g-mobile/products/allip.html.

[13] 3GPP TR 25.897, v0.4.1, Feasibility Study on the Evolution of UTRAN Architecture.

[14] H. Yumiba, K. Imai and M. Yabusaki, "IP-Based IMT Network Platform," *IEEE Personal Communications*, October 2001, pp. 18-23.

[15] "The Benefit of Packet-Switched All-IP Mobile Broadband Network," http://www.flarion.com/technology/All-IP.

[16] M. S. Corson, et al., "A New Paradigm for IP-Based Cellular Networks," *IT Pro*, November/December 2001, pp. 20-29.

[17] Intel, "An Overview of the Modular Communications Platform," http://www.intel.com.

[18] U. Mukherjee, "Designing Next Generation Platforms with AdvancedTCA (ATCA)," *Intel in Communications e-Seminar Series*, May 2003.

[19] PICMG3.0, AdcancedTCA, http://www.picmg.org.

Chapter 8

Autonomic Networks

With the increasing complexity and scale of mobile communications networks, technologies for network management are becoming critical for network operators to control the quality of services and operational expenditures. It is expected that the future mobile communications networks will consist of different generations of cellular networks and other complementary systems such as wireless local area networks (WLANs) and wireless personal area networks (WPANs). The management and maintenance of such heterogeneous networks will be a challenging task. On one hand, the potential benefit of these networks, such as great diversity and huge capacity, cannot be fully exploited without an intelligently coordinated system configuration, performance optimization, and resource management. On the other hand, the extraordinary scale and complexity of such networks mean that the network will be difficult to manage and prone to faults. They may demand highly skilled professionals to carry out fault correction and routine maintenance, thus resulting in high cost.

A promising solution to the above problems is to substantially increase the intelligence level of the network and its elements to enable self-configuration, self-optimization and self-healing, so that the maximum amount of traffic is handled with satisfactory quality of services and minimum human effort and interference are needed. In this book, we call these highly desirable networks autonomic networks. The term autonomic is derived from the body's autonomic nerve system, which controls key functions without a conscious awareness or involvement, and the concept of autonomic network is partly borrowed from the world of computing [1]. An autonomic network should be capable of knowing itself, running itself, adjusting to varying circumstances, and preparing its resources to handle the traffic loads most efficiently. It should be equipped with redundancy in the configurable hardware and with downloadable firmware. When faults happen in the network or when the network is attacked, it should repair the malfunctioning parts and protect itself with minimal or zero human intervention. In such a network, smart algorithms will be used to determine the most efficient and cost-effective way to send and transmit information to its destined users.

In this chapter, the principle of the operation and management (O&M) of radio access networks is discussed and the concept of autonomic networks is presented. It is shown how one can apply artificial intelligence (AI) techniques such as Bayesian belief networks, neural networks, and case based reasoning to achieve the goal of autonomic networks. Furthermore, the concepts of distributed network management employing mobile agents and ant colony optimization (ACO) are discussed.

8.1 O&M OF MOBILE RADIO NETWORKS

The operation and maintenance of mobile radio networks are traditionally divided into several functional areas: configuration management (CM), performance management (PM), fault management (FM), state management (SM), software management, inventory management, and security management. Their scopes can be described in the following sections.

8.1.1 Configuration Management

Configuration management (CM) is responsible for the definition of the configuration, immediate or deferred configuration changes, storage of configuration data, and issuance of warnings about changes that may affect service provision. To be specific, CM deals with information concerning the initial installation, extension, reduction, modification and monitoring of the network elements, and network resources. This functionality ensures that the network operator achieves the optimal system configuration [2].

8.1.2 Performance Management

Performance management (PM) is responsible for defining and collecting performance measurement data from the respective network elements. The measurement data include user and signaling traffic, resource access and availability, and quality of services. This information can be used to identify and eliminate potential bottlenecks, thus ensuring ongoing customer satisfaction.

To evaluate the performance of the network, it is required to collect and record performance data at all network elements according to a schedule established by the element managers, and these data are reported to the network manager for analysis. The current 3GPP specification on performance management is mainly focused on the requirements to produce these data, but not on the analysis [3]. The areas of performance measurement include the following:

- Traffic levels within the network including the level of both the user traffic and the signaling traffic;
- Verification of the network configuration;

- Resource access measurements;
- Quality of services such as delays during call setup and packet throughput;
- Resource availability that includes, for instance, the recording of the begin and end times of service unavailability.

8.1.3 Fault Management

Fault management (FM) provides network surveillance and status monitoring functions. It enables the operator to detect and correct network faults quickly and efficiently. This capability is critical to maintaining high quality of service levels of the network.

Fault management has been specified in detail by 3GPP [4]. It includes the detection of faults in the network and the notification of alarms to the operation systems including element managers (EM) and network managers (NM) (see Section 8.1.7). Depending on the nature of the fault, it may cause a change of the operational state of the logical or physical resources. Detection and notification of these state changes are very critical. A list of active alarms in the network and operational state information as well as alarm/state history data provides the basic information for fault detection and correction. Additionally, test procedures can be used in order to obtain more detailed information if necessary, or to verify an alarm, a state, or the proper operation of network elements (NEs) and their logical and physical resources.

Faults that may occur in the network can be classified as follows:

- Hardware failures such as the malfunction of some physical resource within a NE;
- Software problems such as software bugs and database inconsistencies;
- Functional faults (for instance, a failure of some functional resource in a NE can be found without an apparent hardware component problem associated with it);
- Loss of some or all of the NE's capability due to overload situations;
- Communication failures between two NEs, or between a NE and the operating system (OS), or between two OSs.

As a consequence of faults, appropriate alarms related to the physical or logical resources affected by the faults are generated by the network entities. When any type of fault described above occurs within a 3G network, the affected network entities should be able to detect them immediately. The network entities accomplish this task by using autonomous self-check circuits or procedures, including, in the case of NEs, the observation of measurements, counters, and thresholds. The threshold measurements may be predefined by the manufacturer and executed autonomously in the NE, or they may be based on performance measurements administered by the EM.

The majority of the faults should have well-defined conditions for the declaration of their presence or absence, that is, fault occurrence and fault clearing conditions. Such an incident is referred to as a *steady fault.* The network entities should be able to recognize it if a previously detected steady fault is no longer present. For some faults, no clearing condition exists and they are referred to as *unsteady faults.* An example of this is when the network entity has to restart a software process due to some inconsistencies, and normal operation is resumed afterwards. In this case, although the inconsistencies are cleared, the cause of the problem is not yet corrected. Manual intervention by the system operator will always be necessary to clear unsteady faults since these, by definition, cannot be cleared by the network entity itself.

According to [4], the fault detection process is required to provide at least the following information for each fault:

- For hardware faults, the smallest replaceable unit that is faulty;
- For software faults, the affected software component such as corrupted files, databases, or software code;
- For functional faults, the affected functionality;
- For faults caused by overload, the reason for the overload;
- An indication of the physical and logical resources that are affected by the fault;
- The severity of the fault including indeterminate, warning, minor, major, and critical;
- The probable cause of the fault;
- The nature of the fault, that is, steady or unsteady;
- Any other information that may help understand the cause and the location of the abnormal situation.

8.1.4 State Management

State management (SM) retrieves the state/status attributes of managed objects. It monitors changes of state/status attributes of managed objects and allows the operator to configure state/status attributes for administrative purposes. The state management in UTRAN is defined by ITU for Telecom Network Management [5]. State management offers a means to operators to monitor and control their network resources. State and status attributes in the managed objects provide detailed information about the current condition of network resources. The given states and status attributes offer information about various conditions concerning the availability and operability of the network resources to the operator.

8.1.5 Software Management

Software management is responsible for the import and export of files and the administration of the network entity software including activation, download, upload, and version control.

8.1.6 Inventory Management

Inventory management deals with the collection and storage of identification information for each Field Replaceable Unit (FRU) including hardware, firmware, and resident software configurations for each network element. This ensures that operators know exactly what resides within their networks and helps ensure rapid fault resolution.

8.1.7 Security Management

Security management controls the access to the network management system. It deals with such issues as access permission, passwords, automatic password aging, alarms, and audit trails. Its aim is to reduce the risk of unauthorized access, thus increasing the confidence of network operators and their customers.

8.1.8 3GPP Architecture of Network Management

As an illustration, Figure 8.1 shows the network management reference model specified by 3GPP for 3G systems [6], where the following management interfaces are identified:

- The interface between network elements (NE);
- The interface between the network elements, such as base stations and RNCs, and the element manager (EM) of a single public land mobile communications network (PLMN);
- The interface between the element manager and the network manager (NM) of a single PLMN organization. Note that it is possible that the element manager functionality may reside in the NE, in which case this interface is directly from NE to NM. These management interfaces are given the reference name Itf-N;
- The interface between the network managers of a single PLMN organization;
- The interface between the network managers and the enterprise systems of a single PLMN organization;
- The interface between enterprise systems and network managers of different PLMN organizations.

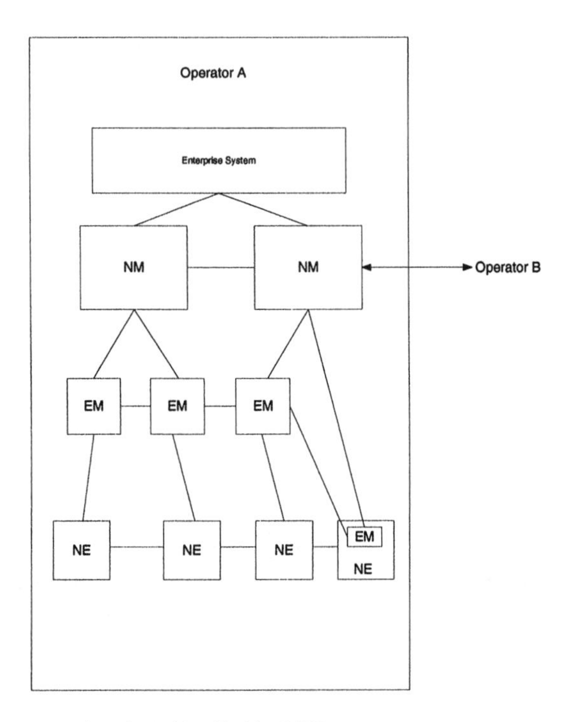

Figure 8.1 Management interfaces identified by 3GPP.

The current 3GPP specification is primarily focused on the interfaces between the element managers and the network manager of a single PLMN organization. It should be noted, however, that the goal of autonomic networks cannot be fully reached without specifying other interfaces.

In UTRAN, the RNC needs to coordinate a large number of base stations (node Bs) and manage the overall radio resources. To this end, the exchange of management information between the RNC and node Bs and also between different RNCs is required. However, this type of information shared between the RNC and NodeB is currently not considered necessary to the network managers and is therefore dealt with separately by a different interface [7]. All other management information related to node B is transparently transferred by the RNC towards the UTRAN network management system (see Figure 8.2). Such an arrangement of interfaces serves the network management of current UTRAN well. According to our vision of autonomic networks, however, a closer cooperation of O&M and radio resource management is needed for future mobile communications networks.

From these discussions, it is seen that a sophisticated network management system has been introduced to UTRAN. However, the focus of the standards is mainly on data collection and reporting. The network intelligence including data analysis and decision-making is largely left to the endeavor of network vendors. In the following sections, it is shown how to increase the level of network intelligence by applying AI techniques.

It should be pointed out that all the O&M functionalities discussed above are indispensable for the management of the mobile radio access networks. To simplify the discussions in this chapter, however, we shall group them according to the three goals of autonomic networks: self-awareness, self-optimization, and self-healing. State management, fault management, configuration management, and performance management all contribute to the first goal, self-awareness. Fault management, configuration management, inventory management and software management contribute to the second goal, self-healing. Configuration management and performance management contribute to the third goal, self-optimization. Admittedly, the relationship between the conventional O&M functionalities of a mobile radio network and autonomic networks is complex and these associations only serve as good examples.

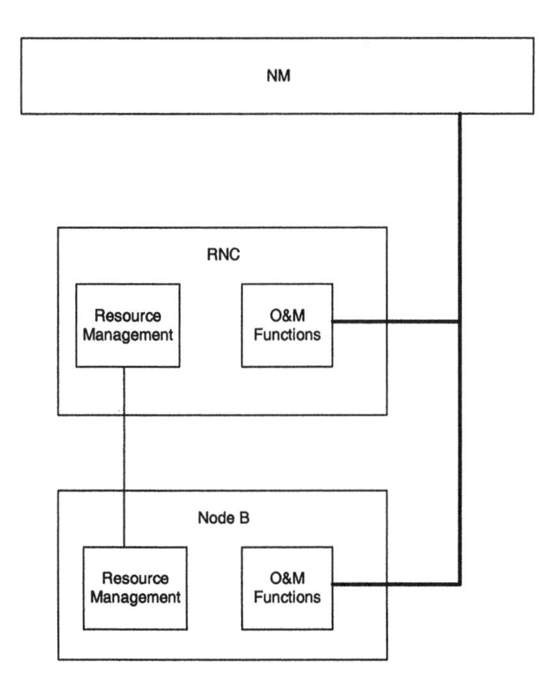

Figure 8.2 An illustration of two distinct interfaces.

8.2 FUNDAMENTALS OF AUTONOMIC NETWORKS

The goal of autonomic networks is to increase the automation and reduce human intervention in the management of networks. At the minimum, an autonomic network should have the following three features. First, an autonomic network should be self-aware. It should have a detailed knowledge of its elements that includes current states of all the network elements, traffic load across the network, and ultimate capacity, internal network topology, and all connections to other networks. It needs to know the extent of its own resources, including the shared ones among network elements and fixed ones dedicated to certain elements. In a higher level, the network entities in an autonomic network must know both themselves and their surrounding networks, including their activities, and should act accordingly. They should follow and update general rules to interact with neighboring network elements in order to achieve global optimization. They will tap available resources and negotiate the use by other network elements of its underutilized resources, configuring both itself and its connections to other networks in the process.

Second, an autonomic network should have the capability of self-configuring and self-optimizing. It should never settle for the status quo; it always looks for ways to improve its performance. It should monitor its constituent parts and change global and local network parameters in order to achieve the desired goals of the network operators. An autonomic network must also configure and reconfigure itself under varying conditions. System configuration or setup should occur automatically, as well as dynamic adjustments to that configuration to best handle changing traffic flows and hardware and software resources.

Third, an autonomic network should be capable of self-healing. It should be able to recover from routine and extraordinary events that might cause some of its parts to malfunction. It should be able to discover faults or potential problems, and then find corrective measures in an automatic manner. These can be activating redundant modules, sending software patches, offloading the processing and traffic load to other network elements, or reconfiguring the system to keep it functioning smoothly. An autonomic network should be an expert in self-protection. It should detect, identify, and protect itself against various types of attacks to maintain overall system security and integrity.

8.3 SELF-OPTIMIZATION

A radio access network consists of a large number of access nodes and network controlling units. In such a network, it is very important to manage the utilization of bandwidth resources both in the air and in the transport network, and to configure the network elements dynamically. There are two aspects in the optimization process: allocation of resources to each connection of mobile terminals and to each IP packet transportation, and configuration and optimization of the network entities. These two types of optimization are conventionally called resource management and system configuration and optimization, respectively, and they are operated at two different levels.

To understand the first aspect, consider the example of connecting a mobile terminal to the mobile communications network. First, given the capability of the mobile terminal, one needs to make the choice of the appropriate su-network. Second, one needs to decide on the specific radio access node to be used. For instance, a mobile terminal may be closer to access node A than access node B. But if access node A is heavily loaded, it may be better to use access node B to realize the maximum network throughput. Third, one needs to choose the best frequency and the best channel type for carrying the information to and from the terminal. The latter includes dedicated channels, shared channels, and common channels. Once packets are sent to the IP-based transport network, one needs to find the best route for them to reach their destinations. Since a mobile communications network needs to handle a huge number of mobile terminals at a given time, the decisions discussed above must be made in accordance with the loading conditions in the network. With given traffic load and performance, the network needs to adjust its global and local parameters, redistribute the network resources and reconfigure the hardware so that the quality of services can be maintained across the whole network, which is the second and the major aspect of self-optimization.

The tasks of the network optimization include both configuration and performance management. The self-optimizing capabilities allow mobile communications networks to autonomously measure the performance or usage of resources and then tune the configuration of hardware resources to deliver improved performance. Some typical optimization questions that a mobile communications network manager has include the following: What is the traffic flow in different parts of the network? Is there any congestion taking place? What is the global picture of the quality of services across the network? How can one reduce delay and increase the throughput and the quality of services? There are several measures that are commonly used in the optimization of the radio access networks. These include system handover, frequency handover, or relocation, power control, channel switching, and changing antenna parameters such as antenna height and tilt.

The classical network optimization process consists of the following steps: information collection, data analysis, configuration change, and verification. In the first step, the configuration data and performance measurements are collected. In the second step, the collected data are analyzed to determine if the network is running at its best and if there are any corrective actions to be taken. In the third step, instructions are sent to the network elements to make the configuration change. Last but not least, a verification test is carried out to ensure that the change of configuration leads to improved performance. In a self-optimizing network, all these tasks should be accomplished automatically and the automation makes it easier and faster to respond to the network dynamics.

Admittedly, self-configuration can be regarded as a feature of its own merit for network management. In this book, however, self-configuration has been included as part of the self-optimizing process because sometimes it is difficult to distinguish one from the other. In a self-optimizing network, new hardware resources are seamlessly integrated with the old ones and configured in a co-ordinated manner to achieve global optimization. Hardware subsystems and resources can configure and reconfigure autonomously both at boot time and during run time. This action may be initiated by the need to adjust the allocation of resources based on the current optimization criteria or in response to hardware or firmware faults. Self-optimization also includes the ability to concurrently add or remove hardware resources in response to commands. An example is when hierarchical cells are deployed. Assume that the resources are initially allocated to provide the right mix for normal operation. If the carrier used by the picocells is not in use due to some unrecoverable hardware failure, for instance, the network can reallocate the carrier and some hardware resources from the pico cell to the macro cell. On the other hand, if the macro cell is faulty, the network can shift out the resources to the overlayed pico cells. This is feasible if the remote RF head and node B farming technologies are employed (see Chapter 2).

Siemens Mobile Adaptive Radio Technology (S.M.A.R.T.) represents the first step towards self-optimizing networks [8]. It offers the network operator a chance to achieve superb quality of services by semi-adaptive network optimization. Currently, most mobile communications network operators in Europe are employing the UTRAN network on the first chosen carrier. With the growth of data-oriented traffic, it is expected that the traffic volume will increase at a great pace in some geographical areas some time in the future. This will result in high capacity demand across the network and thus require more carriers and more base stations. In general, every change of frequency allocation requires a complete reoptimization of a part of the network, which could become too complicated to be handled efficiently with conventional planning tools. Besides, since the frequency band is a limited nature resource, other optimization strategies must be employed to enhance the capacity and the quality of existing networks and increase spectrum efficiency. With S.M.A.R.T, network configurations and optimizations will be fully automated. S.M.A.R.T. will provide valuable information, such as analysis results and recommendations, to both the operator's operation and maintenance (O&M) center and network planning departments. In an advanced stage of S.M.A.R.T. development, it is expected that operators will only need to monitor rather than actively manage the network. S.M.A.R.T. is addressing some key areas of the network operation and maintenance and it covers several important areas including smart carrier allocation, smart interference reduction, automatic traffic pattern analysis, automatic tuning of handover and power control parameters, and automatic radio planning. As a commercial tool, it is expected that S.M.A.R.T. will be introduced to Siemens' customers in a pragmatic manner, with the automation level increasing year by year.

8.4 FAULT MANAGEMENT AND SELF-HEALING

A self-healing network is one that can survive the failure of network entities or links without human intervention. The healing property should be provided at both the physical layer and higher layers. A technique called hot redundancy provides a self-healing property in the physical layer. In this technique, the network nodes are equipped with redundant hardware to receive and process real-time data for key devices or modules. When a device breaks, the redundant one becomes operational instantly without booting and configuring. An IP network provides inherent healing property at the network layer. The failure of a router or leased line in the IP backbone causes the routing protocol to propagate new reachability status throughout the network. All the routers read the information and calculate new routes between endpoints, which they then use for future packet forwarding. With self-healing capabilities, platforms are able to detect hardware and firmware faults instantly and then contain the effects of the faults within defined boundaries. This allows platforms to recover from the negative effects of such faults with minimal or no impact on the execution of the operating system, middleware, and user-level data.

The conventional technology used to perform the basic task of network healing is referred to as fault management and its objective is to detect, isolate, and repair failures in networks [11–13]. The fault management system proactively diagnoses the cause of abnormal network behavior, and proposes and, if possible, takes corrective actions. Basically, network faults can be classified into hardware and software faults. The effects of such faults vary from underperformance and local traffic congestion to network breakdown. Examples of hardware faults include the failure of a device due to errors in its logical design, or elements malfunctioning due to simple wear and tear or through external forces such as accidents, acts of nature, mishandling, and vandalism. Examples of software faults include failure of elements due to incorrect or incomplete design of the software, erratic behavior of elements or the network due to software bugs, and slow or faulty services by the network. The flow of fault management can be described as follows: (1) collect alarms; (2) filter and correlate the alarms; (3) diagnose faults through analysis and testing; (4) determine a plan for correction, display correction options to users, and implement the correction plan; (5) verify that the fault is eliminated; and (6) record data and determine the effectiveness of the current fault management function.

The first step in fault management is to collect the monitoring and performance alarm. Typically, alarms are produced by either local network element managers (EM) or by a statistical analysis of the network that monitors the trends and threshold crossings. Alarms can be classified into two categories, physical and logical. Physical alarms are typically caused by hardware failures, normally reported through an element manager. Logical alarms are caused by malfunctioning of certain network entities and they tend to appear as statistical errors, an example of which is performance degradation due to congestion. The second step in fault management is to filter and correlate the alarms. Alarm filtering is a process that analyzes the multitude of alarms received and eliminates the redundant alarms, such as multiple occurrences of the same alarm. Alarm correlation is the interpretation of multiple alarms in order to derive new conceptual meanings and create new alarms, and to link the multialarms with possible root causes. Faults are identified by such methods as analyzing the filtered and correlated alarms, associating them with past experience and testing. Once a fault is diagnosed, corrective procedures are undertaken by the network to eliminate the cause of the fault. The fault management system's role in correction is to develop a plan or series of actions, and to initiate this plan with other functions within the network. In principle, as much correction as possible should be performed automatically without human intervention, although at times it is necessary for engineers to physically go to a site to replace a part, or for a programmer to debug some software. The correction must be verified through testing requests sent to the element managers, where if the fault does not disappear, more data is analyzed and the diagnostic process is repeated. Another step in fault management is to collect data about the effectiveness of the fault management process in order to improve the process.

There are two types of fault management systems commonly used by network operators. The first one is operator-driven in which the fault diagnosis and corrective measures are carried out manually. The second one is rule-based expert system, in which rules crafted by network vendors are used to diagnose faults and to rectify problems. Faults are normally reported to the network managers by alarms, which can be related with environment, communications, processing error, hardware, and QoS. The root cause of the alarms can be often found through correlation. Although expert systems represent the right move in the direction of self-healing networks, they are not sufficient to meet the need of self-healing in future mobile communications networks.

With the complexity of the heterogeneous future mobile communications networks, more traffic, more nodes, more equipment types, and more protocols will be expected. In order to reduce labor cost and reduce service downtime, maintenance tasks need to be automated. Further, preventive measures should be taken before any serious failure taking place. These tasks call for the application of wide-range artificial intelligence (AI) techniques. Artificial intelligence technology can play an important role in problem solving and reasoning that are fundamental to automated fault management. There are several techniques that have been developed or are being studied in the field of artificial intelligence and, when being integrated properly, they should be able to solve most problems in fault management and help reach the goal of self-healing networks.

8.5 APPLICATION OF ARTIFICIAL INTELLIGENCE

As discussed in the previous section, the complex process of fault management includes alarm filtering and correlation, fault identification, correction, and verification. These functions involve such operations as correlation, pattern recognition, and data interpretation. Consciously or unconsciously, decisions are made by virtue of a knowledge base that contains descriptions of network elements and topology, categorization, and a problem-solving strategy. Artificial intelligence technology is well suited for these kinds of applications. Admittedly, expert systems have been used for fault diagnosis in some networks, which represents the right move of network management towards self-healing. However, expert systems do have some limitations that make them difficult to deal with future mobile communications networks alone. Generally speaking, it is inconvenient for expert systems to handle new, incomplete, and ambiguous data. Rules in expert systems tend to be brittle and not robust when faced with unforeseen situations. An example is a new combination of alarms due to changing network topology. Expert systems cannot learn from experience. To be specific, they cannot use analogy to reason from past experiences or remember past successes and failures in the context of a current problem. The rules that are incorporated at the development stage cannot easily adapt as the network evolves, as it would require accurate and comprehensive prediction of future events. They do not scale well to large, dynamic, real-world domains. It is difficult, especially for technicians or operators not familiar with artificial intelligence and the system, to add new rules without a comprehensive understanding of what the current rule base is and how a new rule may impact the rule base. They are not good at handling probability or uncertainty. These drawbacks argue for the employment of a wide range of artificial intelligence techniques that can overcome the difficulties mentioned above, either alone or as an enhancement of the expert system. Probabilistic methods such as neural networks (NNs) or Bayesian belief networks (BBNs) are appropriate for correlation, while symbolic methods such as case-based reasoning (CBR) or expert systems are appropriate for fault identification [14, 15]. In many cases it is beneficial to use these techniques in cooperation with each other. Another area of fault management where the artificial intelligence technology can have a positive impact is fault correction. CBR systems and expert systems can develop plans or courses of action that will correct a fault that has been identified. The application of these methods and the reasoning behind using them are discussed in the following sections.

8.5.1 Alarm Filtering and Correlation

Alarm filtering is comprised of four processes: compression, count, suppression, and generalization [12, 13]. Compression is the reduction of multiple occurrences of the same alarm into a single alarm; count is the substitution of a specified number of occurrences of similar alarms or alarm categories with a new alarm; suppression is inhibiting a low-priority alarm in the presence of a higher-priority alarm, and generalization refers to an alarm by its superclass where the superclasses are determined by experts. These four processes are well defined and can thus be achieved through the use of rules or the application of expert systems.

In addition to filtering, the alarms need to be correlated. Correlating alarms is a difficult task due to their inherent ambiguity. Even with large detailed amounts of data, there can still be a significant amount of uncertainties and inconsistencies. In many cases, there is more than one plausible explanation for the underlying cause of a group of alarms. For instance, the nonoccurrence of a remote event may cause a device waiting for that event to time out, with several possible causes for this lack of response; the device could be faulty, the response could be delayed due to congestion, or the local device's timer could be faulty. The production of incomplete data due to crash-recovery cycles is another problem area for correlating alarms. Some network elements may have built-in mechanisms to reset themselves when a local fault tolerance level has been reached. This can cause the destruction of important evidence needed to diagnose faults, and can indicate that there is nothing wrong in the system while the failure condition still exists, thus resulting in repetitive crash-recovery cycles. The complex process of correlation can be regarded as substituting a set of alarms that match a predefined pattern with a new alarm. It is basically a pattern recognition problem based on ambiguous, incomplete, and inconsistent data. Therefore, alarm correlation can be automated by employing a probabilistic artificial intelligence technique such as neural networks (NN) or Bayesian belief networks (BBN).

8.5.2 Neural Networks for Alarm Correlation

Feedforward neural networks have already been proven effective in medical diagnosis, target tracking from multiple sensors, and image and data compression [9]. In principle, neural networks can be used as an effective tool for accomplishing alarm correlation. In fact, the following properties of the multilayer feedforward neural network make it a powerful tool.

- Neural networks can recognize conditions similar to previous conditions for which the solution is known.
- They can approximate any function, given enough neurons, including Boolean functions and classifiers. This gives NNs great capability to be trained for different alarm patterns without a deep understanding of the knowledge domain.

- They provide a fast and efficient method for analyzing incoming alarms.
- They can handle incomplete, ambiguous, and imperfect data.

A feedforward neural network is shown in Figure 8.3, where the neurons are arranged into three layers. In this model, there is a single input layer, a single output layer, and zero, one, or more hidden layers in between. As the name suggests, all connections are in the forward direction and there is no feedback. The source nodes in the input layer of the network supply respective elements of the activation pattern, which constitutes the input signals applied to the neurons in the second layer. The output of the second layer is used as the input signal for neurons in the third layer, and this process is continued to the rest of the network. The output of the signals of the neurons in the output layer of the network constitutes the overall response of the network to the activation signal supplied to the source nodes.

Feedforward networks are useful because of their ability to approximate any function provided that sufficient neurons are employed, and their ability to learn from samples of input-output pairs. Learning is accomplished by adjusting the connection weights in response to input-output pairs, and training can be done either offline or online during actual use. Depending on how the training is done, these neural networks (NNs) can be characterized as being trained by supervised methods or by unsupervised methods. Supervised NNs training data consists of correct input vector/output vector pairs used to adjust the neural network connection weights. An input vector is applied to the NN, the output vector obtained from the NN is compared with the correct output vector, and the connection weights are adjusted to minimize the difference. A well-trained neural network can successfully generalize what it has learned from the training set. Given an input vector not in the training set, the NN produces a correct output vector most of the time. In unsupervised training, there is no training data based on known input/output pairs. The NN discovers patterns, regularities, correlations, or categories in the input data and sorts them in the output. For example, an unsupervised neural network where the variance of the output is minimized could serve as a categorizer which sorts inputs into various groups.

Unsupervised training is typically faster than supervised training and it provides the possibility of discovering new patterns. For these reasons unsupervised training should be used even in situations where supervised training is possible. However, for alarm correlation, the fact that the input/output pairs can be easily produced makes supervised trained NNs a plausible choice.

8.5.3 Bayesian Belief Networks for Alarm Correlation

Bayesian belief networks (BBN) are another possible choice for alarm correlation due to their ability to handle uncertainty and represent cause and consequence relationships [10]. A Bayesian belief network defines a set of events, the dependencies between them, and the conditional probabilities associated with the dependencies. By virtue of the mathematics of Bayesian probability, a Bayesian belief network can be used to calculate the probabilities of variable causes being the actual cause of a given event.

Bayesian networks combine two mathematical approaches, Bayesian statistics and graph theory, to provide a powerful means of modeling probabilities based on continuously updated information. Using nodes and arcs to represent uncertain events and their connections, respectively, a BBN is a representation of cause-and -effect dependencies. The information known about one node (i.e., effect node) depends on the information of its predecessor nodes that represent its causes. This relationship is expressed by a probability distribution for each effect node, based on the possible values of its predecessor nodes' variables. An effect node can also lead into other nodes, where it then plays the role of a cause node. It should be noted that, in a BBN, it is not necessary to build huge joint probability distribution tables that include permutations of all the nodes in the network. Instead, only a node's immediate predecessor's possible states and their effects on the node are needed. Owing to the BBN's form of knowledge representation, large amounts of interconnected and causally linked data can be represented. Furthermore, using Bayesian belief networks, programs can dynamically learn by constantly modifying probabilities using a fixed set of rules.

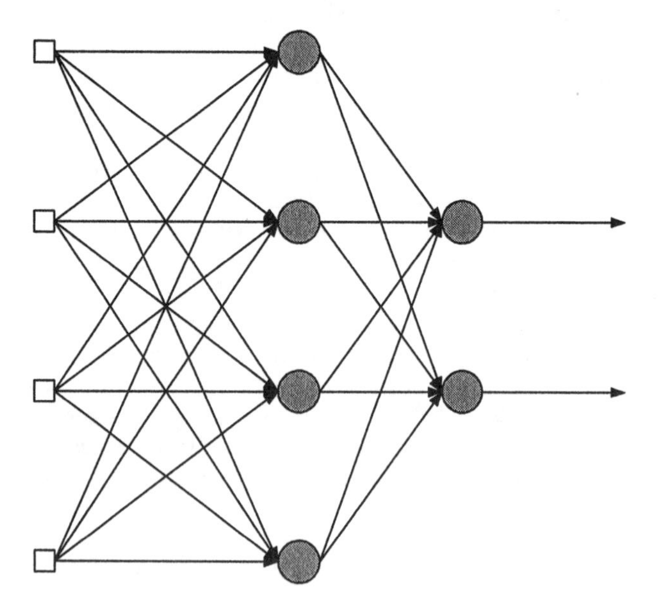

Figure 8.3 Illustration of a feedforward neural network.

Generally speaking, BBNs have the following features:

- They can represent deep knowledge by modeling the functionality of the transmission network in terms of cause-and-effect relationships between element and network behavior and faults.
- They can provide guidance in diagnosis. Calculations over the same BBN can determine both the precedence of alarms and the areas that need further clarification in order to provide a finer-grained diagnosis.
- They can handle noisy, transient, and ambiguous data due to the fact that they are based on probability theory.
- Compared to other probabilistic methods, they have a modular, compact, and easy-to-understand representation.

BBNs are appropriate for automated diagnosis because of their deep representations and precise calculations. A concise and direct way to represent a system's diagnostic model is to use a BBN constructed from relationships between failure symptoms and underlying problems. A BBN represents cause and effect between observable symptoms and the unobserved problems so that when a set of symptoms is observed, the problems most likely to be the cause can be determined. In practice, the network is built from descriptions of the likely effects for a chosen fault. When used as a diagnostic tool, the system reasons from effects back to causes.

An excellent example of using BBN as a tool for trouble shooting is the Microsoft Bayesian Network (MSBN) [11]. In MSBN, information is organized as problems, causes, evidence, and resolutions. Problems (alarms) are at the top of the hierarchy. Under each problem, there is a list of all of the possible causes. Under each cause, there is a list of all of the evidence that identifies a cause. Finally, for each cause, there are one or more solutions. If a cause appears under more than one problem, the evidence and resolutions need to be the same for all appearances of the cause. For each cause, there are typical problems as consequences and typical evidences to observe. It is understood that MSBN has contributed to the great success of Microsoft operating systems.

The development of a diagnostic BBN requires a deep understanding of the cause-and-effect relationships in the given discipline, provided by experts. This is both an advantage and a disadvantage. The advantage is that the knowledge is not represented as a black box but as an NN. Thus, humanly understandable explanations of diagnoses can be given. The disadvantage is that the realm of future mobile communications networks is technologically evolving and the topology, architecture, and types of network entities are all to be defined, and the knowledge needed for fault analysis is not fully available. Therefore, the BBN network itself needs to be fully adaptive.

8.5.4 Fault Identification with Case-Based Reasoning

The filtering and correlation of alarms are the first step of fault diagnosis. The second step involves further analysis and identification of the root cause of the alarms, the fault. The third step is to find the correct measure to remove the fault. The second and third steps are closely related and they naturally form an iterative process; after the alarm data are analyzed, decisions are made whether more data should be gathered, a finer-grained analysis should be executed, or problem-solving should be performed. Gathering more data can consist of sending tests to network elements or requesting network performance data. Problem solving requires expert knowledge about the network elements, topology, and typical faults. A symbolic artificial intelligence approach such as expert systems or case-based reasoning (CBR) is ideal for such applications [14, 15].

Case-based reasoning is a relatively new problem-solving paradigm in artificial intelligence, which utilizes the specific knowledge of previously experienced and concrete problem situations. It is based on the assumption that situations recur with regularity. Studies of experts and their problem-solving techniques have found that experts rely strongly on applying their previous experiences to the current problem at hand. CBR can be regarded as such: an expert that applies previous experiences stored as cases in a case library. Thus, the problem-solving process becomes one of recalling old experiences and interpreting the new situation in terms of those old experiences. A new problem is solved by finding a similar old case and reusing it in the new problem situation. It is an approach of incremental and sustained learning. New experience is retained every time a problem is solved and the new case is made available for future problems. Major tasks of the cased-based learning process include identification of the current problem, finding a past case similar to the new one, and using that case to suggest a solution, evaluation of the proposed solution, and update of the system by learning from this experience.

Figure 8.4 illustrates the life cycle of CBR systems for problem-solving. It is shown that there are four steps in the process: (1) case retrieval; (2) case interpretation and adaptation; (3) solution evaluation and modification; and (4) implementation and learning. The first step is retrieving cases that best match the current situation or case. To this end, it is crucial to use an appropriate indexing method, such as decision trees or nearest neighbor matching. Once a case is retrieved, it must be interpreted and adapted. The interpretation process is a simple comparison between the retrieved cases and the current case. Adaptation is a complicated, domain-dependent process that uses rules to adapt the current case to the problem situation and proposes an initial solution, based on the similarities and differences. The next step is an evaluation and modification cycle where the proposed solution is evaluated through comparisons to cases with similar solutions or through simulation, and the solution is modified accordingly. After the CBR system has found its best solution, the solution is implemented and the results are evaluated. The resulting evaluation, solution steps, and problem context are entered into a new case, which is then indexed into the case library, thus allowing the system to learn.

Compared with expert systems, CBR systems have the following advantages:

- They can handle new and changing data through their ability to use analogy.
- They can learn from experience through the acquisition of new cases.
- They can scale well to large knowledge domains due to the ability of their knowledge representation structures to collapse or merge into each other.
- They do not require extensive maintenance of the knowledge base.
- They can judge the "goodness" of a proposed solution based on simulations and/or previous cases from the case library.
- They can use a method of knowledge acquisition that is less time-consuming than the rule development of an expert system.

In case-based reasoning, experiences are contextualized pieces of knowledge that are stored as cases in a case library. An important component of any CBR system is to decide what the key attributes of a case are and on which attributes the cases will be indexed. Indices are combinations of the cases' attributes or descriptors and are used to predict which cases are useful to the current situation. Typically, the frontend of an application that uses CBR will have a graphical user interface (GUI) that allows a user to easily input new cases without requiring any detailed understanding of the underlying CBR engine. The cases can be input in text and are automatically indexed into the case library.

8.6 A HYBRID AI APPROACH TO SELF-HEALING NETWORKS

Undoubtedly, self-healing is a highly desirable feature for the management of the future large and dynamic mobile communications networks. With self-healing, the fault management system should be able to adjust to changes that occur within the network elements and topology and recover from faulty operations automatically. To this end, a neural network should be trained or a Bayesian-based network be built to filter, correlate, and identify general categories of faults. In order to identify the exact location of a fault, a procedure is also necessary where appropriate tests are requested and the results analyzed to further pinpoint a fault's location. However, this procedure involves a series of decisions based on human expertise and cannot be implemented by an NN or a BBN alone. Therefore, it is appropriate to train a NN or construct a BBN to correlate alarms and recognize basic fault patterns and to use symbolic processing such as CBR to further analyze the data, run tests, and identify and locate the fault.

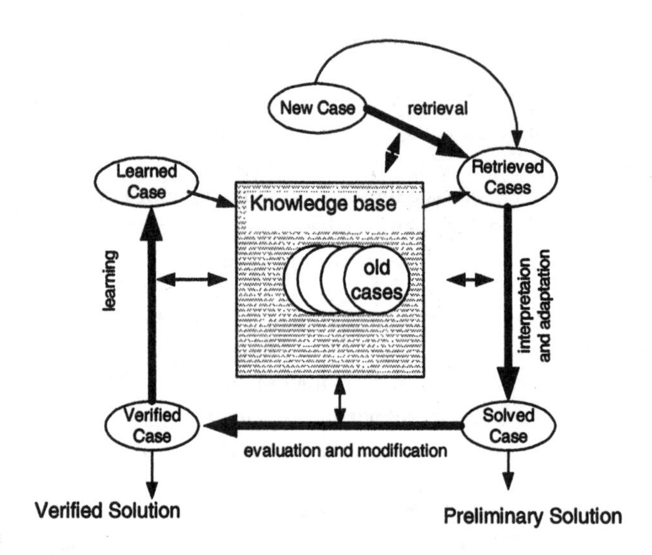

Figure 8.4 The life cycle of a case-based reasoning system.

A hybrid artificial intelligence system that uses both probabilistic and symbolic problem-solving AI techniques is an ideal approach to realizing the self-healing networks due to the diverse nature of the fault management task [15]. Instead of performing the whole task with one technique that has the strength only in some aspects, the hybrid system employs different techniques to suit the specific tasks. Thus, the strength of each technique is made use of to the full and its weakness is compensated for by the other. A drawback in using a hybrid artificial intelligence system, however, is that knowledge acquisition must be performed more than once and in very different ways. For example, a neural network must be trained with large amounts of input/output data pairs and a CBR system must be seeded with initial cases drawn from experts and other symbolic data sources. This adds time to the development process and requires two types of knowledge representations. However, the advantages of robustness and accuracy achieved using a hybrid system in fault management can outweigh the costs.

An example of a hybrid AI system for self-healing networks is shown in Figure 8.5. In the figure, network alarms are fed into an expert system that filters the alarms through compression, count, suppression, and generalization. Filtered alarms serve as input into a neural network, and the neural network is trained to recognize common fault categories. The output of the neural network consists of the most likely fault types, protocol types, and geographic area of the fault. This analysis is then fed into a case-based reasoning system where faults are diagnosed, corrective measures are suggested and verification is performed. The CBR system can make decisions, based on past problem solving experiences, whether to gather more data, run the data through a different NN that was trained on a finer-grained set of alarms, or implement the solutions.

8.7 DISTRIBUTED NETWORK MANAGEMENT

A further possible move to reach the goal of autonomic networks is to distribute the intelligence across the network [17]. An intelligent distributed network should possess the following features. First, the network should operate with or without a centralized command node. Second, each node should either have the knowledge of the overall network or know how to behave in order to achieve a global optimum status, or both. Third, every node should collaborate with its neighboring nodes. Fourth, every node should have the adaptive learning ability to develop its intelligence through individual and collective experiences.

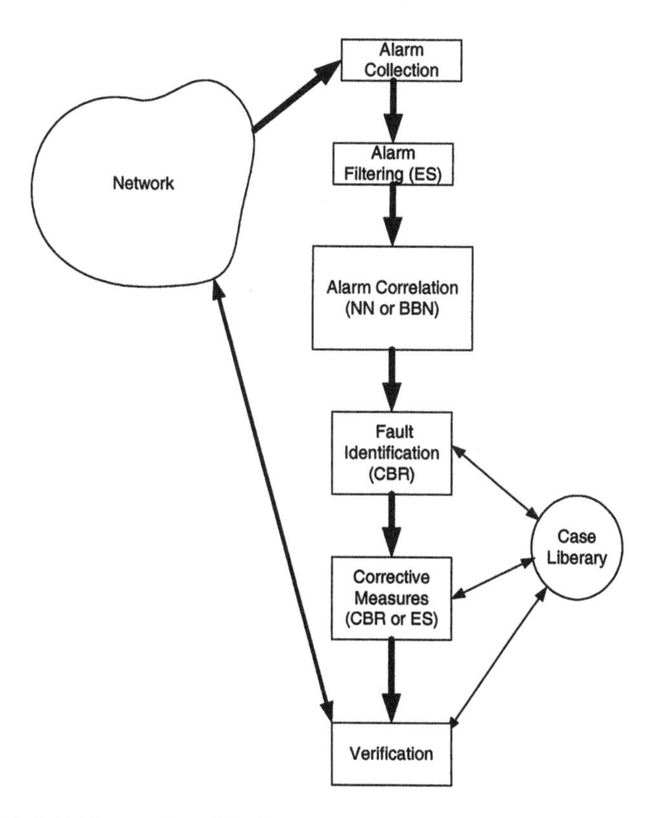

Figure 8.5 A hybrid AI system for self-healing networks.

A technique for realizing autonomic networks with distributed intelligence is to use intelligent mobile software agents [18, 19]. Intelligent software agents are independent executable programs which are capable of acting autonomously in the presence of expected and unexpected events. In a multiagent system, a group of autonomous agents interact with the environment and with each other in a coordinated manner to achieve certain goals. When an agent is equipped with the ability to move from one node to another to accomplish its task, it is called a mobile agent. For large and complex mobile communications networks, the use of mobile intelligent multiagents could find problems at different network elements and solve them quickly before they propagate and accumulate.

Although the distributed network management may sound like an over-ambitious approach for managing the whole mobile communications networks in the immediate future, especially in the areas of fault management, the concept is at least valid for accomplishing some tasks. A problem well suited for the intelligent mobile multiagents is routing. The application of mobile agents to congestion control in a circuit switched environment was reported in [20]. The work was funded by British Telecom Labs and it demonstrated the feasibility rather than the practicability of using agents to autonomously manage a network. The argumentation was that adopting a mobile agent strategy could lead to intrinsic robustness to system and program failure. In [20], a large number of homogeneous agents were used to control routing problems in a simulated synchronous digital hierarchy (SDH) network with 30 nodes. The type of agents used were "Telescript" mobile agents. The control was provided by using an "ensemble of mobile agents" of two different kinds. The "ensemble" came from the researchers' requirements that for robust control agents should be present in reasonably large numbers. In fact, the number of agents used to control the network dynamically changed in order to adapt to the dynamics of the network state. The two different kinds of agents described were *load agents* to provide the lowest level of control in the system and *parent* agents. The *load agents* were responsible for distributing the communications traffic evenly through the available circuits. The parent agent's tasks were to manage the population dynamics of the *load agents*. The *parent agent* possessed the ability to "launch" new mobile *load agents* when it detected nodes that were in high states of utilization. In the experiments reported, the mobile load agents were allowed to move across the network to measure the nodal utilization, which is defined as the amount of traffic throughput in a single network node. The load agents traveled in a backward manner, starting from destination node towards the source node, generating a localized knowledge of the network node utilization states. This behavior was achieved via the use of a distributed implementation of Dijkstra's shortest path algorithm. The load agents generated the best available routes with the lowest total node utilization. The mobile load agents were able to obtain a global view of the network if they were allowed to traverse all of the network nodes. Unfortunately, it was found later that this method of node searching has the drawback that sometimes cyclic routes were generated [21].

A different approach to the mobile agent technology was adopted in [22]. Instead of focusing on moving software code from node to node, an ant-based intelligence was used. A multitude of simple homogeneous software code, which was referred to as ants, were allowed to randomly roam the physical network. The concept was based on the following observation.

"Watch a group of ants as they try to bring a few drops of sugar solution to their nest. First, a single ant takes a sip and wanders off. Shortly afterwards, dozens of them begin to line up back and forth, following what seems like the shortest route between the nest and the food. It looks as though some high intelligence is at work, but ants have only a few hundred neurons to help them make decisions. In fact, ants do not follow a plan and they just react. The ants' success as foragers stems in large part from using the world as a prompt. When an ant stumbles across food, it does not remember where it is, but it lays scent trails to and from it. To find food, other ants follow these trails. At first, ants choose between a long and short path at random, but because more ants travel the shorter path in a given time, the scent builds up more quickly here. This becomes the favored path. This method of using the world as a memory bank is called stigmergy. "

The behavior of ant colonies and how they coordinate complex activities like foraging and nest building have fascinated researchers in animal behavior. They have proposed many models to explain these capabilities. Recently, algorithms taking inspiration from the behavior of real ant colonies have been applied to solve many types of optimization problems. This new approach to distributed optimization is known as ant colony optimization (ACO). Ant colony optimization has been applied successfully to a large number of combinatorial problems. The classification of ants as a very simple mobile agent system is justified because although the ants in [21] were not directly involved in carrying out control changes in the network, their presence did indirectly influence the control of the network routing strategy.

The work reported in [22] was focused on using the ant analogy for the control of route path finding in networks via stigmergy. The idea was to use the the agents' environment as a memory store. The type of routing considered in this work was hop-by-hop routing, where each node has a routing table which gives information about the next nearest neighbor nodes. The routing table in its most basic form consists of a table of node identifiers each with an associated priority. To route a call, information is given to the source node, which then checks for a node that is consistent with getting to the designated node. Once a node is chosen, the destination information is passed onto the chosen node and the process continues until the destination node is reached. In the research, the simulated ants randomly roamed the network laying a trail of "pheromones" at each node. The amount of pheromone deposited depended on the amount of time each of the ants were delayed in the nodes that they traversed and were directly dependent on the utilization of the node. As the ants aged, the amount of pheromone that they deposited decreased. The amount of pheromone became, in effect, the goodness measure of that route. Further ants followed the routes where higher levels of pheromones had been deposited, therefore reenforcing that path. Paths with high levels of pheromone indicated a less congested route, because congested nodes would delay the ants and therefore prevent the ants from depositing a greater amount of pheromone. In the simulated network, the routing tables were replaced with pheromone tables which list scores that the ants use to navigate. The ants were created constantly at different nodes around the network and traveled to random destinations, where they died. As they traversed the network, they left a scent trail behind them for traffic to follow. As an ant passed through a node, it increased the score in the pheromone table for its source node. The amount that the ant added to the score decreased as the ant got older, so the whole system responds more strongly to ants that found shorter routes. Also, the system delayed ants that passed through congested nodes, thereby biasing the system in favor of ants that found uncongested routes.

8.8 SIMPLE NETWORK MANAGEMENT PROTOCOL

To implement the concept of autonomous networks, the management nodes need to communicate with the other network nodes quickly and reliably. Since the future mobile communications networks are expected to be based on the Internet protocol (IP), a promising protocol for such communications is the simple network management protocol (SNMP) [16].

Since its creation in 1988 as a short-term solution to manage network elements in the Internet, SNMP has achieved wide acceptance as the most popular protocol for the management of the Internet. SNMP is based on the manager and agent model consisting of a manager, an agent, a database of management information, managed objects, and the network protocol. In any configuration, at least one manager node needs to run SNMP management software. Network devices to be managed, such as bridges, routers, and servers, should be equipped with an agent software module. The agent provides the interface between the manager and the physical devices being managed. The manager and agent use a management information base (MIB) and a relatively small set of commands to exchange information. The agent is responsible for providing access to a local MIB of objects that reflects the resources and activity at its node. The agent also responds to manager commands to retrieve values from the MIB and to set values in the MIB. The MIB is organized in a tree structure with individual variables, such as point status or description, being represented as leaves on the branches. A long numeric tag or object identifier (OID) is used to distinguish each variable uniquely in the MIB and in SNMP messages.

In SNMP version 1, five basic messages are used to communicate between the manager and the agent, which include the following: GET, GET-NEXT, GET-RESPONSE, SET, and TRAP. The GET and GET-NEXT messages allow the manager to request information for a specific variable. The agent, upon receiving a GET or GET-NEXT message, will issue a GET-RESPONSE message to the manager with either the information requested or an error indication regarding why the request cannot be processed. A SET message allows the manager to request a change to be made to the value of a specific variable. The agent will then respond with a GET-RESPONSE message indicating the change has been made or an error showing why the change cannot be made. Finally, the TRAP message allows the agent to spontaneously inform the manager of an important event. To increase the efficiency of SNMP, the GET-BULK command was introduced to SNMP version 2 in 1993. This is used to retrieve a large amount of management information in order to minimize the number of protocol exchanges. In addition, other new commands were also included in SNMP version 2, such as INFORM used for communications between managers and REPORT for communications between SNMP engines.

It can be observed that most of the messages including GET, GET-NEXT, and SET are only issued by the SNMP manager. Because the TRAP message is the only message capable of being initiated by an agent, it should be used to report alarms. This notifies the SNMP manager as soon as an alarm condition occurs, instead of waiting for the SNMP manager to ask. The small number of commands used is only one of the reasons SNMP is simple. The other simplifying factor is its reliance on an unsupervised or connectionless communication link. This simplicity has led directly to its widespread use, specifically in the Internet Network Management Framework. Within this framework, SNMP is considered robust because of the independence of the managers from the agents. For instance, should an agent fail, the manager will continue to function, or vice versa.

Each SNMP element manages specific objects with each object having specific characteristics. Each object has a unique object identifier (OID) consisting of numbers separated by decimal points, for instance, 1.2.3.4.5.1.2682.1. These object identifiers naturally form a tree. The MIB associates each OID with a readable label and various other parameters related to the object. The MIB then serves as a data dictionary or code book that is used to assemble and interpret SNMP messages. When an SNMP manager wants to know the value of an object, such as the state of an alarm point, the system name, or the element uptime, it will assemble a GET packet that includes the OID for each object of interest. The element receives the request and looks up each OID in its MIB. If the OID is found, that is, the object is managed by the element, a response packet is assembled and sent with the current value of the object included. If the OID is not found, a special error response is sent that identifies the unmanaged object. When an element sends a TRAP packet, it can include OID and value information (bindings) to clarify the event. The manager sends a GET or GET-

NEXT to read variables and the agent's response contains the requested information. The manager sends a SET to change variables and the agent's response confirms the change if it is allowed. The agent sends a TRAP when a specific event occurs.

An SNMP message is not sent by itself. It is wrapped in the User Datagram Protocol (UDP), which in turn is wrapped in the Internet protocol (IP). With reference to the four layer model for the Internet, SNMP resides in the application layer, UDP resides in the transport layer, and IP resides in the Internet layer. The fourth layer is the network interface layer where the assembled packet is actually interfaced to some kind of transport media (i.e., twisted pair copper, coaxial or fiber). As an example, assume that a SNMP manager sends a GET request to the agent. The SNMP manager wants to know what the agent's system name is and prepares a GET message for the appropriate OID. It then passes the message to the UDP layer. The UDP layer adds a data block that identifies the manager port to which the response packet should be sent and the port on which it expects the SNMP agent to be listening for messages. The packet thus formed is then passed to the IP layer. Here a data block containing the IP and media access addresses of the manager and the agent is added before the entire assembled packet gets passed to the network interface layer. The network interface layer verifies media access and availability and places the packet on the media for transport. After working its way across bridges and through routers based on the IP information, the packet finally arrives at the agent. Here it passes through the same four layers in exactly the opposite order as it did at the manager. First, it is pulled off the media by the network interface layer. After confirming that the packet is intact and valid, the network interface layer simply passes it to the IP layer. The IP layer verifies the media access and IP address and passes it on to the UDP layer where the target port is checked for connected applications. If an application is listening at the target port, the packet is passed to the application layer. If the listening application is the SNMP agent, the GET request is processed. The agent's response then follows the identical path in reverse to reach the manager.

It should be pointed out that neither SNMP version 1 nor SNMP version 2 offers security features. To be specific, SNMP version 1 and version 2 can neither authenticate the source of a management message nor provide encryption. Without authentication, it is possible for nonauthorized users to exercise SNMP network management functions. It is also possible for nonauthorized users to eavesdrop on the exchanged information as it passes from managed systems to the management system. To rectify these security problems, SNMP version 3 was issued in 1998. This set of documents for SNMP version 3 does not provide a complete SNMP capability but rather defines an overall SNMP architecture and a set of security capabilities. These are intended to be used with the existing SNMP version 2. Simply put, SNMP version 3 is SNMP version 2 plus administration and security. SNMP version 3 includes three important services: *authentication, privacy,* and *access control.* To deliver these services in a flexible and efficient manner, SNMP version 3 employed the concept of a *principal*, which is the entity on whose behalf services are provided or processing takes place. A principal can be an individual acting in a particular role; a set of individuals with each acting in a particular role; an application or set of applications; or combinations thereof. In essence, a principal operates from a management station and issues SNMP commands to agent systems. The identity of the principal and the target agent together determine the security features that will be invoked, including authentication, privacy, and access control. The use of principals allows security policies to be tailored to the specific principal, agent, and information exchange and gives human security managers considerable flexibility in assigning network authorization to users [16].

8.9 CONCLUDING REMARKS

Autonomic networks are aimed at making networks self-managing, thus shifting the burden of network management to technologies. This will not only reduce the cost of network operation and maintenance, but also increase the network efficiency and quality of services. In our vision, autonomic networks rely on the application of artificial intelligence technology in software engineering and reconfigurability of the hardware. Notwithstanding, the path to automatic networks is not straightforward. First, smart and efficient artificial intelligence algorithms are needed to run on powerful processors to cope with varying situations in a timely manner. Second, open industrial standards are needed to ensure interoperatability of different subsystems and different products, so information such as alarms and states of network entities can be exchanged. Third, the benefit of providing automation must exceed the cost, which is difficult to achieve for some network entities. For instance, with the current state of the art, providing redundancy for some radio devices is too costly.

It should be pointed out, however, that autonomic networks should be regarded as the pursuit of technology, and the actual implementation must be pragmatic. For instance, the intelligence can be introduced to a network from network elements, to subnetwork level and then to the whole network. Also, one can introduce intelligence to some local or partial management functionality first and then extend it globally. Regardless of the detailed evolution path, the autonomic networks will be realized in future mobile communications networks.

References

[1] J. O. Kephart and D. M. Chess, "The Vision of Autonomic Computing," *IEEE Computer Magazine*, January 2003.

[2] 3GPP TS 32.106: "3G Configuration Management."

[3] 3GPP TS 32.104 v4.0, "Telecommunication Management, 3G Performance Management (PM)."

[4] 3GPP TS 32.111 v3.0.1, "3G Fault Management."

[5] ITU-T Recommendation X.731: Information Technology–Open Systems Interconnection –Systems Management: State Management Function.

[6] 3GPP TS 32.101: 3G Telecom Management Principles and High Level Requirements.

[7] 3GPP TS 32.102: 3G Telecom Management Architecture.

[8] http://www.siemens.com.

[9] S. Haykin, *Neural Networks*, New York: Macmillan, Inc., 1994.

[10] S. Russell and P. Norvig, *Artificial Intelligence: A Modern Approach*, Upper Saddle River, NJ: Prentice Hall, 1995.

[11] J. Locke, "Basic Knowledge Engineering," December 1999, http://freelock.com/technical/bays.php.

[12] G. Jakobson and M. D. Weissman, "Alarm Correlation," *IEEE Network Magzine*, November 1993, pp. 52–59.

[13] S. Aidarous and T. Plevyak (eds.), *Telecommunications Network Management into the 21st Century*, New York: IEEE Press, 1994.

[14] A. Aamodt and E. Plaza, "Case-Based Reasoning: Fundamental Issues, Methodological Variations and System Approaches," *AICOM-Artificial Intelligence Communications*, IOS Press, Vol. 7, pp. 39-59.

[15] D. W. Gürer, et al., *An Artificial Intelligence Approach to Network Fault Management*, Menlo Park, CA: SRI International.

[16] W. Stallings, *SNMP, SNMPv2, SNMPv3, and RMON 1 and 2*, 2nd ed., Reading, MA: Addison Wesley, 1998.

[17] D. Gurer, A. Sastry, and V. Lakshminarayan, "Adaptive Management of Heterogeneous Networks by Using Distributed Artificial Intelligence," *EXPERSYS*'98, Virginia Beach, VA, November 1998.

[18] A. L. G. Hayzelden and R. A. Bourne, *Agent Technology for Communications Infrastructure*, New York: John Wiley and Sons, 2001.

[19] L. G. Hayzelden and J. Bigham, "Software Agents in Communications Network Management: An Overview," *Knowledge Engineering Review*, 1999.

[20] S. Appleby and S. Steward, "Software Agents for Control," in P. Cochrane and P. Mealthey, *Modelling Future Telecommunication Systems*, New York: Chapman & Hall, 1994.

[21] B. Schoonderwoewrd, O. Holland, and J. Bruten, "Ant-Like Agents for Load Balancing in Telecommunications Networks," *Proceedings of the First International Conference on Autonomous Agents*, 1997, pp. 209-216.

[22] M. Ward, "There Is an Ant in My Phone," *New Scientist*, January 24, 1998.

Chapter 9

Ubiquitous Networks

In the preceding chapters, we have discussed emerging technologies for future radio access networks from the perspective of vendors and operators. These technologies will enhance the capacity of mobile communications networks, increase the user data rate and quality of services over the air, and reduce the operation and maintenance expenditure for mobile network operators. From a user's perspective, however, the ultimate goal of mobile and wireless communications should be to provide a ubiquitous information network [1]. The word "ubiquitous" originates from a Latin word, meaning "present in all places at the same time."

In the early days of data communications, networking between computers was done by connecting a number of terminals to a costly mainframe to share the information and processing power within an organization. In the 1990s, with the emergence of the Internet and the wide deployment of mobile communications networks, came the era of personal computing and personal communications in which individuals could exchange information by personal computers and mobile phones. Today we are entering the era of converged communications and computing in which both mobile phones and personal computing devices are becoming integrated. To some extent, every communications device computes and every computing device communicates.

In the future, it is expected that the mobile communications infrastructure will be ubiquitous, and different types of networks will be fully integrated. The networks for voice and data will converge to a common data network based on IP. Mobile terminals will become networked information terminals that include all or part of the functionalities of mobile phones, computers, personal digital assistants (PDA), digital television, digital video recorder, and game consoles. The information terminals will also offer the functionality of intelligent sensors that support people's daily lives both at home and at work. This will be done by a large group of devices placed at various locations in the environment where they automatically sense conditions in the real world, exchange the information gathered, take actions in coordination and send the collective information to a remote controller when needed. In Weiser's vision of ubiquitous computing, these terminals should be pervasive and calm, receding to the background so that people will be unaware of their existence [2, 3].

In this chapter, the requirement of ubiquitous networks on the network infrastructure is discussed and the convergence of different types of mobile communications networks is elaborated. In particular, we shall examine two parallel approaches to network convergence, which are being taken by the cellular industry and the IEEE 802 working groups. These two examples of convergence illustrate a much larger universe of activities that moves forward on many planes and fronts in the realm of access networks.

9.1 REQUIREMENT ON THE NETWORK

To realize the vision of ubiquitous networks, the future network infrastructure must provide the information terminals simple and high-speed connection to broadband networks anywhere and at any time [1, 3]. Although it is not possible to predict accurately the infrastructure needed for ubiquitous networks, there are some key elements that are required to achieve it. The first element is the integration of different radio access networks including both cellular networks and other short-range wireless networks. We call this the convergence of mobile radio access networks and it will be discussed in detail in Section 9.2. The second element is the integration of fixed and mobile communications networks. It is foreseeable that future mobile information terminals will not necessarily access the telecommunications networks via a conventional radio access network. When convenient, a mobile terminal may get connected to a fixed network through a short-distance wireless system, such as wireless local area networks (WLAN) and wireless personal area networks (WPAN). When the mobile terminal moves out of the coverage area of the short distance wireless network, it can be handed over to a cellular network. The third element is digitalized broadcasting and spectrum sharing (or dynamic spectrum resource allocation) among different broadcasters and mobile network operators in order to ease the problem of radio resource shortage [4]. The wonderful aspect of broadcasting is that the information from a single broadcast can be shared simultaneously by any number of people. The problem we have at the moment is that broadcasters all enjoy their own dedicated frequency bands. Owing to the relatively low technology employed by some broadcasters and the lack of cooperation between different service providers, the overall spectrum efficiency for broadcasting is low. On the other hand, the cost for mobile network operators to broadcast and multicast multimedia information is high. Fortunately, we are beginning to see the appearance of digital audio broadcast (DAB) and digital video broadcast (DVB) services. With the advent of software radio technology used in various forms of mobile terminals, it makes technological and economical sense for broadcasters and mobile operators to share spectrum resources. In fact, the cooperation of mobile operators and broadcasters will enable the former to have much greater downlink capacity and the latter to have an uplink that is needed for interactive and personalized services. In Asia, there are some encouraging signs of broadcasters working together with cellular operators to offer such services. This naturally leads to the fourth key element, broadband wireless communications. Wireless access with broad bandwidth is essential if we are to enjoy a variety of promising new services. This has been demonstrated by the popularity of broadband access to the Internet.

Mobile radio access technologies can play at least three vital roles in ubiquitous networks. The first one is to provide a broadband high capacity mobile network with global reach so that individual mobile terminals can be connected to the global Internet anywhere and communicate any information. The second one is the enabling of wireless local area networks and wireless personal area networks based on technologies such as IEEE 802 systems. This will extend the geographical reach of mobile communications networks and increase the data rate over the air in hot-spot and industrial, office, and home environments. The third one is to make mobile mini-networks connected to either a cellular network or a fixed wireless network via a mobile terminal. In this case, the mobile terminal may serve the roles of a router, an access node, and an IP-enabled mobile terminal. The beauty of such configuration is that the mini network can be a mobile one, such as a personal or a vehicle network. It requires a new entity within the mobile terminal to provide the functionalities of the router and server. Such an entity is known as the personal mobile gateway (PMG) [5].

In our opinion, the personal mobile gate way of the U.S. firm IXI [5] represents the right technology advance towards ubiquitous networks. IXI's PMG is aimed at integrating cellular and short-distance wireless with micro-router and micro-server functionalities. It is the point of connection between the mobile radio access network and the new category of smart devices equipped with "ubiquitous communicators" such as watches, pens, phones, messaging terminals, gaming devices, and cameras. It can be a stand-alone device of the size of a small mint box, or be integrated into a legacy mobile terminal. Personal gateways can bring additional revenues to mobile operators and manufacturers by enabling new services to mobile network subscribers. It gives consumers the freedom to add to their collection of connected devices according to their budget, over time, and by using a single subscription to the mobile network operator, thus enriching their mobile experience in a pragmatic manner.

9.2 THE CONVERGENCE OF MOBILE NETWORKS

In order to understand how mobile radio access networks may converge in the future, it is instrumental to examine the path taken by the cellular industry, called the 3G path, and that taken by IEEE 802 working groups.

9.2.1 The 3G Path

Up until now, most cellular networks including both the second and part of the third generation systems, are circuit-switched systems based on the SS7 signaling protocol. These SS7-based networks are fully optimized for mobile users in terms of authentication, authorization, and accounting (AAA), mobility management, and service provisioning. Since these kinds of access networks are voice-centric, they are being augmented by packet-based services such as GPRS and EDGE for GSM, 1xEV-DO and 1xEV-DV for CDMA2000, and HSDPA for UTRAN. In these networks, unfortunately, the subscribers' packets have to cross many media borders, so the quality of services (QoS) can fall far short of customer expectations. Also, it is rather expensive to maintain two separate networks for voice and data. Currently, the 3GPP networks are evolving in the direction of all IP and the integration with high-speed short-distance IP based networks such as wireless LANs (WLANs). The goal in the foreseeable future is to provide ubiquitous networks using IP as the sole networking technology and using unified authentication, authorization, and accounting systems controlled by cellular operators.

Figure 9.1 shows the architecture of a de facto WLAN system [6], in which WLAN terminals access the network via the access point (AP). Different access points are connected by a layer 2 distribution network, which is connected to the IP backbone network via an access router (AR). The user profile is stored in the user database (DB) and the authentication, authorization, and accounting functions are served by the AAA server. The AAA server is typically based on the RADIUS protocol, a popular one used in Internet access networks.

Cellular network operators have large customer bases, proven security, respectable customer care records, and reliable charging and billing systems. Therefore, the current 3GPP-WLAN interworking group is aimed at serving WLAN users with the UTRA AAA server and home subscriber servers (HSS) (see Figure 9.2) [6, 7]. When a WLAN terminal accesses the WLAN network, it can choose its home cellular network among a list of virtual public land mobile networks (PLMNs) that have roaming agreement with the WLAN operator. This integration approach has the advantage of protecting 3G operators' investment. Since UTRA networks do provide IP access over the packet-switched core network (PS CN) domain, it is cost effective and can be easily extended to the integration of cellular networks with other IP-based data networks such as wireless personal area networks (WPANs). Unfortunately, although this approach offers an integrated billing service and subscriber profiles, it does not guarantee session continuity across network boundaries between WLAN and UTRAN.

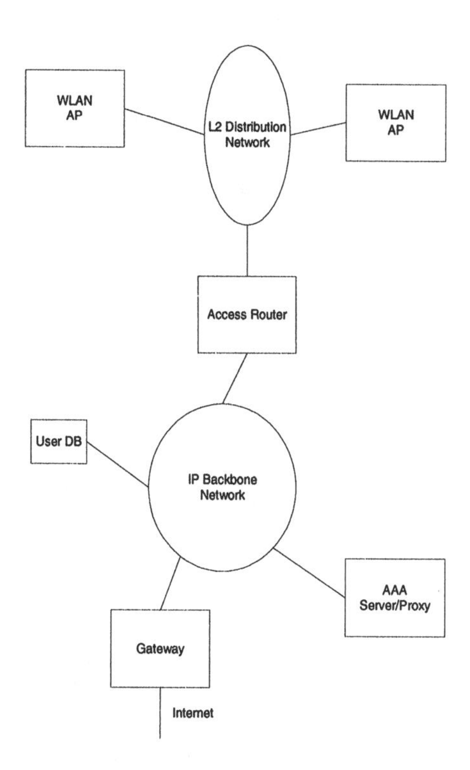

Figure 9.1 A de facto WLAN system architecture.

An alternative 3G-WLAN interworking approach is proposed in [8] for CDMA2000 1x, which employs mobile IP to enable seamless handover between networks to preserve ongoing sessions. As discussed in Chapter 7, the CDMA2000 1x core network does support mobile IP. Therefore, it can be directly integrated with a WLAN network via a WLAN gateway (see Figure 9.3). The WLAN gateway is required to support mobile IP to handle mobility and AAA services to interwork with the AAA servers in the CDMA2000 home network. This will enable CDMA2000 1x operators to collect the WLAN accounting records and generate unified billing statements for subscribers.

The above methods for 3G-WLAN interworking are called "loosely coupled," as the WLAN networks can be directly connected to the Internet [8]. Another way of integrating 3G networks with WLANs is called "tightly coupled interworking." Its concept is to make the WLAN appear to the 3G core network as another radio access network. In this case, the WLAN gateway hides the details of the WLAN network to the 3G core, and implements all the 3G protocols required in a 3G radio access network such as mobility management and authentication. Mobile terminals in this approach are required to implement the corresponding 3G protocol stack on top of the standard WLAN protocol stack, and switch from one physical layer to the other if needed. In effect, all the traffic generated by clients in the WLAN network is injected into the 3G core network using 3G protocols. The disadvantages of this approach are that it requires significant changes to the WLAN terminals and needs explicit physical connection to the CDMA2000 1x core network. Therefore, the loosely coupled interworking is preferred.

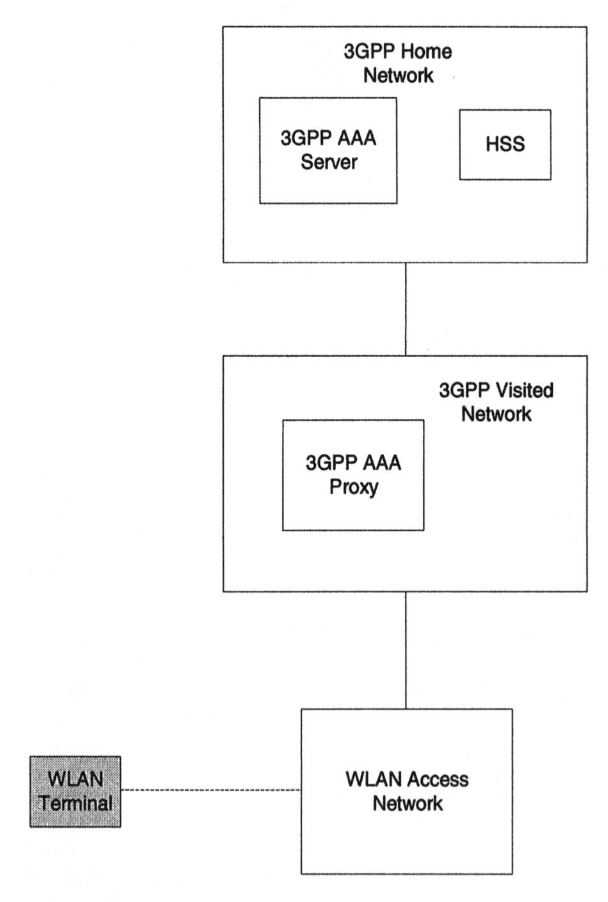

Figure 9.2 3GPP-WLAN interworking architecture.

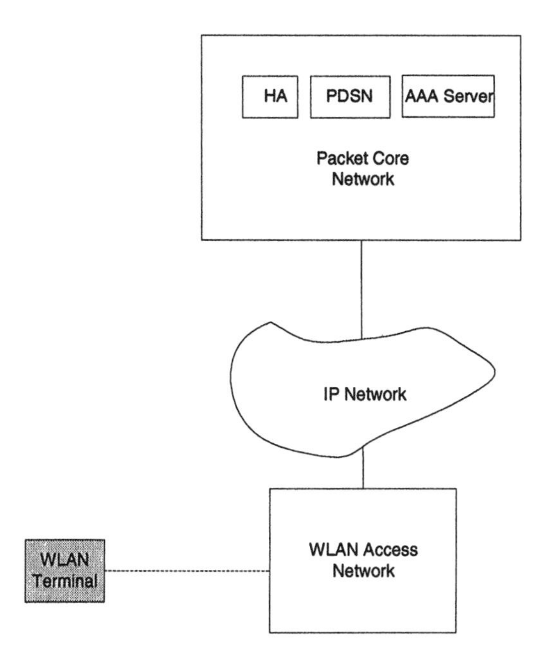

Figure 9.3 An illustration of CDMA2000 1x interworking with WLAN.

9.2.2 The IEEE 802 Path

In contrast to the path taken by 3GPP and most cellular operators in Europe and Japan, a direct all-IP approach to network convergence is being taken by the IEEE 802 working groups [9]. This approach does not try to protect the investment in or depend on the existing circuit-switched infrastructure. Instead, it tries to build ubiquitous wireless extensions to the Internet.

Within the IEEE 802 working groups, a family of wireless systems have been developed to cater to the needs of different systems of wireless communications including wireless local area networks (WLAN, IEEE 802.11) and wireless personal area networks (WPAN, IEEE 802.15). Recently, the IEEE 802 family was expanded to include fixed broadband wireless access (FBWA, IEEE 802.16), wireless metropolitan area networks (WMAN, IEEE 802.16.a), mobile broadband wireless access (MBWN, IEEE 802.20), and the interworking of all the 802 family systems (IEEE 802.21). These systems serve different applications and offer all the component networks needed for ubiquitous networks (see Figure 9.4).

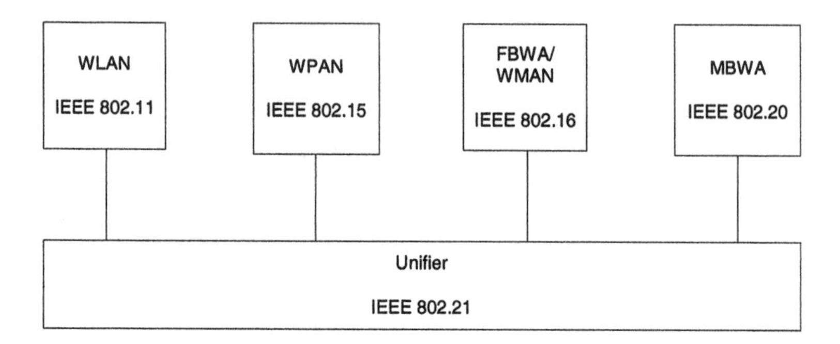

Figure 9.4 The landscape of IEEE 802 systems for wireless communications.

9.2.2.1 IEEE 802.11

The IEEE 802.11 wireless LAN family of standards are the most comprehensive and most influential. It is perhaps the only one beginning to enjoy commercial success. The 802.11 family of systems share two unlicensed frequency bands at 2.4 GHz and 5 GHz. As such, any network complying with the standards and possibly also with some local regulations can be operated without a radio license. Among its ever increasing number of variants, the major ones in the IEEE 802.11 family are the following:

- IEEE 802.11 is the original wireless LAN standard. Developed in 1997, it had a 2 Mbps maximum data rate over the air interface. Two versions of 802.11 were specified, a frequency hopping spread spectrum (FHSS) version and a direct sequence spread spectrum (DSSS) version, both in the 2.4 GHz part of the spectrum. Now, this standard has been largely superseded by IEEE 802.11b.
- IEEE 802.11a operates in the far less congested 5 GHz part of the RF spectrum. Its great advantage is that, by employing a new air interface based on orthogonal frequency division multiplexing (OFDM), it offers a headline data rate of 54 Mbps. In practice, however, most users probably can only share around 20-Mbps bandwidth.
- IEEE 802.11b is the most commercially successful wireless LAN standard. It took the basic DSSS mode of 802.11 and enhanced it with a data rate up to 11 Mbps, although in practice most users can only enjoy around one-fifth of this. The demand on high-speed hot-spot services and the availability of affordable PMCIA cards have resulted in its worldwide adoption.
- IEEE 802.11c provides mechanisms to bridge wireless LANs together to form a single network. Using the 802.11c standard between multiple access points running across a conventional wired network, multiple access points can coordinate their operation, thus allowing members

attached to different access points to exchange data. This technology is meant to be incorporated into various access point products.

- IEEE 802.11d is aimed to adapt the generic IEEE 802.11 standards to different countries and regions to comply with local regulations. In the United States, for example, wireless LAN devices are permitted to operate at up to 1-W transmission power. In the United Kingdom and Europe, however, the maximum transmission power is only 100-mW, thus significantly restricting the range and performance of wireless LANs. The 11d standard was drafted to ensure products were produced that conformed to the local license conditions in each country and region.
- While these IEEE 802.11 standards family can support voice and video communications, the quality of services can rapidly deteriorate if the network is heavily loaded. IEEE 802.11e introduced the concept of prioritization to allow time-sensitive information such as voice and video data to meet the stringent delay requirement.
- IEEE 802.11g introduced the same modulation and air interface schemes as IEEE 802.11a, but in the 2.4-GHz ISM band. Thus, 2.4-GHz WLANs can now enjoy the same data rates as 802.11a. Although it is backwards compatible with IEEE 802.11b, there is some concern about the achievable data rates in mixed 802.11b and 802.11g networks.

From the introduction of wireless LANs, there has been serious concern on their security. IEEE 802.11i introduces stronger security mechanisms to wireless LANs, in which port-based security mechanisms and significantly more advanced and stronger Advanced Encryption Standard (AES) cryptographic algorithms are employed. It provides much greater levels of protection than the previous standard, wired-equivalent privacy (WEP).

9.2.2.2 IEEE 802.15

The IEEE 802.15 specifies a number of systems for wireless personal area networks (WPAN) [10–12]. WPAN devices are defined as those that are carried, worn, or located near the human body. Specific examples of WPAN devices include traditionally networked ones, such as computers, personal digital assistants (PDAs), handheld personal computers, and printers. Also included in WPAN devices are other ones such as digital imaging systems, microphones, speakers, headsets, displays, barcode readers, and sensors. IEEE 802.15 has defined three classes of WPANs that are differentiated by data rate, battery drain, and QoS:

- IEEE 802.15.3 specifies a high data rate WPAN suitable for multimedia communications with very high QoS.
- IEEE 802.15.1 specifies a medium rate WPAN with medium QoS. It is compatible with Bluetooth.
- IEEE 802.15.4 specifies a low rate and low QoS WPAN.

The purpose of IEEE 802.15.3 is to provide a specification for low complexity, low cost, low power consumption, and high data rate wireless WPAN devices. The data rate must be high enough (at least 110 Mbps) to satisfy multimedia and industrial communications. The standard also addresses the quality of service (QoS) capabilities required to support multimedia data types. Products compliant with this standard will complement, not compete with, products compliant with IEEE 802.11 WLANs. The difference between the two is similar to that in the wired world of Ethernet and USB, which provide for connectivity to the network and to peripheral devices, respectively. The 802.15.3 protocol specification requires one of the stations to act as a coordinator, providing the basic timing to the network and managing QoS requirements. The standard also defines a portal as a logical point that integrates the WPAN to a wired or a wireless LAN.

The IEEE working group, 802.15.3a, is aimed at employing ultra-wide bandwidth (UWB) technologies for the air interface of high data rate WPAN. Its technical requirements are summarized in Table 9.1. There are two modes of operation: 110 Mbps and 200 Mbps for different operating ranges, and higher bit rates are possible for shorter operating ranges. The system should be able to operate effectively in the presence of other 802.15.3a systems and other IEEE systems such as WLANs. It is also important that the power consumption be low in order to maintain wireless connectivity on battery-operated portable devices.

Table 9.1

Requirements of 802.15.3a

Parameter	*Lower rate*	*Higher rate*
Bit rate	110 Mbps	200 Mbps
Range	30 ft	12 ft
Power consumption	100 mW	250 mW
Bit error rate	10^{-5}	10^{-5}

According to the FCC, a UWB signal needs either to have more than 20% fractional bandwidth or to occupy more than 500 MHz of spectrum. This means that a UWB signal does not need to be a very short impulse occupying the whole spectrum at the same time. This definition allows the use of multiple bands to encode information in parallel. In a multiband system, information is independently encoded in different bands, instead of being transmitted in a single impulse. This process results in very high bit rate systems with relatively low signaling rates. A multiband UWB system design in general yields many tangible benefits including the following:

- More scalable and adaptive than single band systems;

- Better coexistence characteristics with systems such as 802.11a;
- Lower implementation risks as it leverages more traditional radio design techniques.

These advantages can be retained while maintaining similar complexity and power consumption levels as those of single band systems.

In 2002, IEEE licensed wireless technology from the Bluetooth SIG, Inc., to adapt the Bluetooth specification as base material for IEEE Standard 802.15.1. The approved IEEE 802.15.1 standard is fully compatible with the Bluetooth v1.1 specification. Bluetooth technology defines specifications for small-form-factor, low-cost wireless radio communications among notebook computers, personal digital assistants, mobile phones and other portable, handheld devices, and connectivity to the Internet. The new standard gives the Bluetooth specification greater validity and support in the market.

IEEE 802.15.4 specifies low cost, low power, low rate (250 kbps or lower) and low QoS WPAN systems. A salient feature of this system is that the battery life is on the order of months or years. At the physical layer, it employs the direct sequence spread spectrum (DSSS) technology with 1-MHz chip rate. It uses offset QPSK (O-QPSK) constant modulation scheme in order to reduce the cost of power amplifiers (PA), and orthogonal coding scheme to achieve maximum range. There are 16 channels specified in the unlicensed 2.4-GHz band with 5-MHz channel separation in order to ease the requirement on RF filters. In addition, the IEEE 802.15.4 system can also operate with five channels in the U.S. 915-MHz band and one channel in the European 866-MHz band. To promote the adoption of IEEE 802.15.4 standards, an industrial alliance, ZigBee, has been formed [10]. The ZigBee alliance expects that ZigBee chips will be ubiquitous in smart homes by 2008.

With the wireless PAN technology, future ubiquitous networks will connect not only people, but also objects. With such devices, one can place sensors in jewelry, arm bands, and shirts to form a body network, and the body network communicates with the outside world via a mobile terminal. The sensors will collect the information about the person's physical conditions, including heartbeat rate, blood pressure, and body temperature. When an abnormal situation happens, the mobile terminal will send the collective information to emergency health centers to notify health workers. Another good example is a mini-vehicle network consisting of a personal information terminal and a car navigation system. If the car navigation system can read the driver's schedule stored in his information terminal and understand the traffic information received by the information terminal, it will direct the driver to the right route at the right time. If the driver is late for his or her appointment, a text message can be sent to the other party automatically.

9.2.2.3 IEEE 802.16

The original IEEE 802.16 standard specified the fixed broadband wireless access (FBWA) at frequencies between 10 to 66 GHz. This system provides a communications path between a subscriber site and a core network that is the network to which 802.16 is providing access. IEEE 802.16 standards are concerned with the air interface between a subscriber's transceiver station (mini base station) and a base transceiver station (base station). As the name FBWA suggests, IEEE 802.16 is not aimed to serve mobile subscribers directly. Instead, it can serve as a broadband wireless backhaul for homes and offices, or a distribution and concentration system for separated local area networks.

The IEEE 802.16 employs the time-division multiplexing (TDM) structure. For transmission from subscribers to a base station, the standard uses the demand assignment multiple access-time division multiple access (DAMA-TDMA) technique. DAMA is a capacity assignment technique that adapts according to demand changes among multiple stations. TDMA is the technique of dividing time on a channel into a sequence of frames, each consisting of a number of slots. With DAMA-TDMA, the assignment of slots to channels varies dynamically.

A more important member of the IEEE 802.16 family is the 802.16a system [8]. IEEE 802.16a standard specifies wireless metropolitan area networks (WMANs) in the 2- and 11-GHz band and it defines three physical layers for services: a single-carrier access method which was retained for special-purpose networks; a 256-carrier orthogonal frequency division multiplexed (OFDM) multicarrier for mainstream applications; and a special OFDM standard with 2,048 carriers, which can be used for selective multicast applications, and advanced multiplexing options in tiered metro networks.

At the moment, the most interesting WMAN group is perhaps the IEEE 802.16e, which seeks to increase the level of mobility in WMANs. Compared with those standards for cellular networks, IEEE 802.16e is not designed for handling high-speed handover in a vehicular environment. Instead, it is aimed at low-speed, lightly mobile users who want to maintain some level of roaming between different metro access points. Wireless MANs now are supported by an industrial alliance named the WiMax Forum, which includes many major players in the semiconductor and telecom industry. In areas where no wired infrastructure is in place, WMAN is a viable last-mile solution, and for WLAN hot spots, WMAN is appropriate for backhauls.

9.2.2.4 IEEE 802.20

In December 2002, the IEEE Standards Board approved the establishment of IEEE 802.20, the mobile broadband wireless access (MBWA) working group. The mission of IEEE 802.20 is to develop the specification for an efficient packet based air interface that is optimized for the transport of IP-based services. The goal is to enable worldwide deployment of affordable, ubiquitous, always-on and interoperable multivendor mobile broadband wireless access networks that meet the needs of business and private subscribers. The MBWA standard will specify physical and medium access control layers of an air interface for interoperable mobile broadband wireless access systems, operating in licensed bands below 3.5 GHz, optimized for IP-data transport, with peak data rates per user in excess of 1 Mbps. It supports various vehicular mobility classes up to 250 km/hr in a metropolitan area network (MAN) environment and targets spectral efficiencies, sustained user data rates and numbers of active users that are all significantly higher than achieved by existing mobile systems. In principle, this specification will fill the performance gap between the high data rate, low mobility services currently developed in IEEE 802, and the high mobility cellular systems. The MBWA system can also be used to carry voice over IP (VoIP) traffic as data. Since IEEE 802.20 is designed to be an all IP-data, low-cost, wireless cellular network, however, it could turn out to be a disruptive technology to the cellular industry.

9.2.2.5 IEEE 802.21

All the IEEE 802 systems discussed earlier operate with different air interfaces and target at different market segments. To enable their interworking, a new IEEE working group 802.21 has been established in order to standardize the way session handovers take place between heterogeneous types of IEEE 802-based networks. Its goal is to eventually enable client devices to automatically choose the best available radio network and to seamlessly hand over sessions among networks during roaming without user involvement. This capability will become increasingly important as mobile networks begin supporting real-time applications such as voice, conferencing, remote monitoring, and video transport, which cannot tolerate any perceptible session interruption. Without it, the IEEE 802 family systems would not talk to each other and would not meet the needs of ubiquitous networks.

Internetwork roaming will be achieved by standardizing how IEEE 802 networks pass important network information to the session layer, where handover actually takes place. From an IEEE 802 perspective, there are no procedures for seamless handover yet. The work in IEEE 802.21 is aimed at creating a generic mechanism that works for any IEEE 802 network. Such networks, for example, include the suite of 802.11 wireless LANs, emerging 802.16e wireless MANs and even IEEE 802.3 wired Ethernet LANs. Using mobile IP for roaming today, there are problems with network discovery and timing that cause breaks that can last several seconds. This is because mobile IP as yet has no way to gather relevant information about lower-layer network conditions from IEEE 802 systems.

There are two primary types of information that need to be communicated from Layer 2 (IEEE 802-based networks) to mobile IP or other Layer 3 roaming protocols to enable both fast and seamless roaming among dissimilar network types:

- *Network detection and selection.* When there exist a number of of IEEE 802 systems, the mobile terminal must be able to identify the type of network available, determine whether it can authenticate to it, and find out how much the connection costs. If equipment such as access points and base stations provides this information, client devices can make good handover decisions in selecting the most appropriate network. Currently, this type of information is not shared even between IEEE 802.11 access points.
- *Layer 2 triggers.* This involves Layer 2 providing Layer 3 (IP) with useful information about network conditions. This information could be the degradation of a wireless channel, for example, indicating that a link might fail.

It is important to note that companies the cellular industry are also being represented in IEEE 802.21. This will facilitate the 3G-WLAN interworking as discussed in Section 9.2.1.

9.3 CONCLUDING REMARKS

Undoubtedly, the ultimate goal of mobile radio access technologies is to provide ubiquitous networks to mobile subscribers. Ubiquitous networks require the convergence of different mobile radio access networks including cellular networks, wireless local area networks (WLANs) and wireless personal networks (WPANs). Currently, there are two different approaches to network convergence which are taken by different interest groups. The first one, taken primarily by cellular network operators and vendors, is based on the 3G infrastructure and is aimed at extending the packet switched network (PS) to incorporate other short distance radio technologies such as WLANs and WPANs. Eventually, the data network will provide voice over IP (VoIP) services. The advantage of this approach is that it is pragmatic, so past investment can be partly protected. Also, the quality of voice services can be easily maintained. The second one, taken by IEEE 802 working groups, is trying to directly build a ubiquitous wireless extension to the Internet. The advantage of this approach is that, as it is based on IP, it is suited for high-speed data communications. However, as simple and attractive as they are, they still lack many of the mechanisms that sustain cellular services.

Given the global reach of mobile cellular networks, it is most likely that the 3G path will dominate the evolution of radio access networks. It is expected that some of the IEEE 802 systems will become the wireless extensions of 3G networks. A very interesting member of the IEEE 802 family is the IEEE 802.20 for MBWA. In theory, it could be used as an alternative to 3G cellular networks. In practice, however, it probably serves as a potential major candidate for the fourth generation cellular networks.

Various wireless and mobile communications systems are becoming pervasive in our environment. Ubiquitous networks will become a reality.

References

[1] Ubiquitous Networks, http://www.itu.int/psg/spu/newslog/categories/ubiquitousnetworks.

[2] M. Weiser, "Ubiquitous Computing," http://www.ubiq.com/hypertext/weiser/ubiHome.html.

[3] *IEEE Pervasive Computing Magazine.*

[4] R. Keller et al., "Convergence of Cellular and Broadcast Networks from a Multiradio Perspective," *IEEE Personal Communications Magazine*, Vol. 8, No. 2, pp. 51-56.

[5] http://www.pmgmag.com.

[6] K. Ahmavaara, H. Haverinen, and R. Pichna, "Interworking Architecture Between 3GPP and WLAN Systems," *IEEE Communications Magazine*, Vol. 41, No. 11, November 2003, pp. 74-81.

[7] 3GPP TS 23.234, "Group Services and System Aspects; 3GPP Systesm to Wireless Local Area Network (WLAN) Interworking; System Description (Release 6)," June 2003.

[8] M. Buddhikot et al., "Design and Implementation of a WLAN/CDMA2000 Interworking Architecture," *IEEE Communications Magazine*, November 2003.

[9] http://www.ieee.802.org.

[10] http://www.ieee802.org/15/pub/TG4.html.

[11] http://www.zigbee.org.

[12] http://www.techonline.com/pdf/pavillions/standards/dtc_wpan.pdf.

About the Author

Y. Jay Guo is the manager of system strategy at Mobisphere Ltd., England, which is a Siemens and NEC company for the joint development of the third generation (3G) mobile communications networks. Since joining Mobisphere in September 2000, Dr. Guo has been responsible for the advanced development and future technology. In particular, he has made significant contributions to the development of the high-speed downlink packet access (HSDPA) technology and other cutting edge projects.

Previously, Dr. Guo led an R&D group on advanced technologies for UTRAN base stations at Fujitsu Europe Telecom R&D Centre in England. The projects on which he worked included smart antennas, interference cancelers, and high data rate packet transmission. He also served as Fujitsu's representative in the UTRAN standardization body, 3GPP RAN, and in the steering committee of the U.K. Virtual Centre of Excellence in Mobile Communications (Mobile VCE).

Prior to joining Fujitsu, Dr. Guo was a senior research fellow at the University of Bradford, England, where he supervised a sizable research group on CDMA cellular systems, wireless local area networks (WLANs), signal processing, and antennas and propagation.

Dr. Guo has over 20 years of industrial and academic experience, in both England and China. He holds more than 10 patents in the areas of mobile and wireless communications. He has published two technical books and more than 80 papers in top research journals and at international conferences. He often serves as an expert assessor to various national and international research councils such as the British Engineering and Physical Sciences Research Council (EPSRC) and the Australian Research Council (ARC).

Dr. Guo holds a B.Sc. in electrical engineering, an M.S. in RF engineering, and a Ph.D. in antennas and propagation. He is a senior member of the IEEE.

Index

P

Q

R

Recent Titles in the Artech House Mobile Communications Series

John Walker, Series Editor

3G CDMA2000 Wireless System Engineering, Samuel C. Yang

3G Multimedia Network Services, Accounting, and User Profiles, Freddy Ghys, Marcel Mampaey, Michel Smouts, and Arto Vaaraniemi

Advances in 3G Enhanced Technologies for Wireless Communications, Jiangzhou Wang and Tung-Sang Ng, editors

Advances in Mobile Information Systems, John Walker, editor

Advances in Mobile Radio Access Networks, Y. Jay Guo

CDMA for Wireless Personal Communications, Ramjee Prasad

CDMA Mobile Radio Design, John B. Groe and Lawrence E. Larson

CDMA RF System Engineering, Samuel C. Yang

CDMA Systems Engineering Handbook, Jhong S. Lee and Leonard E. Miller

Cell Planning for Wireless Communications, Manuel F. Cátedra and Jesús Pérez-Arriaga

Cellular Communications: Worldwide Market Development, Garry A. Garrard

Cellular Mobile Systems Engineering, Saleh Faruque

The Complete Wireless Communications Professional: A Guide for Engineers and Managers, William Webb

EDGE for Mobile Internet, Emmanuel Seurre, Patrick Savelli, and Pierre-Jean Pietri

Emerging Public Safety Wireless Communication Systems, Robert I. Desourdis, Jr., et al.

The Future of Wireless Communications, William Webb

GPRS for Mobile Internet, Emmanuel Seurre, Patrick Savelli, and Pierre-Jean Pietri

GPRS: Gateway to Third Generation Mobile Networks, Gunnar Heine and Holger Sagkob

GSM and Personal Communications Handbook, Siegmund M. Redl, Matthias K. Weber, and Malcolm W. Oliphant

GSM Networks: Protocols, Terminology, and Implementation, Gunnar Heine

GSM System Engineering, Asha Mehrotra

Handbook of Land-Mobile Radio System Coverage, Garry C. Hess

Handbook of Mobile Radio Networks, Sami Tabbane

High-Speed Wireless ATM and LANs, Benny Bing

Interference Analysis and Reduction for Wireless Systems, Peter Stavroulakis

Introduction to 3G Mobile Communications, Second Edition, Juha Korhonen

Introduction to Digital Professional Mobile Radio, Hans-Peter A. Ketterling

Introduction to GPS: The Global Positioning System, Ahmed El-Rabbany

An Introduction to GSM, Siegmund M. Redl, Matthias K. Weber, and Malcolm W. Oliphant

Introduction to Mobile Communications Engineering, José M. Hernando and F. Pérez-Fontán

Introduction to Radio Propagation for Fixed and Mobile Communications, John Doble

Introduction to Wireless Local Loop, Second Edition: Broadband and Narrowband Systems, William Webb

IS-136 TDMA Technology, Economics, and Services, Lawrence Harte, Adrian Smith, and Charles A. Jacobs

Location Management and Routing in Mobile Wireless Networks, Amitava Mukherjee, Somprakash Bandyopadhyay, and Debashis Saha

Mobile Data Communications Systems, Peter Wong and David Britland

Mobile IP Technology for M-Business, Mark Norris

Mobile Satellite Communications, Shingo Ohmori, Hiromitsu Wakana, and Seiichiro Kawase

Mobile Telecommunications Standards: GSM, UMTS, TETRA, and ERMES, Rudi Bekkers

Mobile Telecommunications: Standards, Regulation, and Applications, Rudi Bekkers and Jan Smits

Multiantenna Digital Radio Transmission, Massimiliano "Max" Martone

Multipath Phenomena in Cellular Networks, Nathan Blaunstein and Jørgen Bach Andersen

Multiuser Detection in CDMA Mobile Terminals, Piero Castoldi

Personal Wireless Communication with DECT and PWT, John Phillips and Gerard Mac Namee

Practical Wireless Data Modem Design, Jonathon Y. C. Cheah

Prime Codes with Applications to CDMA Optical and Wireless Networks, Guu-Chang Yang and Wing C. Kwong

QoS in Integrated 3G Networks, Robert Lloyd-Evans

Radio Engineering for Wireless Communication and Sensor Applications, Antti V. Räisänen and Arto Lehto

Radio Propagation in Cellular Networks, Nathan Blaunstein

Radio Resource Management for Wireless Networks, Jens Zander and Seong-Lyun Kim

RDS: The Radio Data System, Dietmar Kopitz and Bev Marks

Resource Allocation in Hierarchical Cellular Systems, Lauro Ortigoza-Guerrero and A. Hamid Aghvami

RF and Microwave Circuit Design for Wireless Communications, Lawrence E. Larson, editor

Sample Rate Conversion in Software Configurable Radios, Tim Hentschel

Signal Processing Applications in CDMA Communications, Hui Liu

Software Defined Radio for 3G, Paul Burns

Spread Spectrum CDMA Systems for Wireless Communications, Savo G. Glisic and Branka Vucetic

Third Generation Wireless Systems, Volume 1: Post-Shannon Signal Architectures, George M. Calhoun

Traffic Analysis and Design of Wireless IP Networks, Toni Janevski

Transmission Systems Design Handbook for Wireless Networks, Harvey Lehpamer

UMTS and Mobile Computing, Alexander Joseph Huber and Josef Franz Huber

Understanding Cellular Radio, William Webb

Understanding Digital PCS: The TDMA Standard, Cameron Kelly Coursey

Understanding GPS: Principles and Applications, Elliott D. Kaplan, editor

Understanding WAP: Wireless Applications, Devices, and Services, Marcel van der Heijden and Marcus Taylor, editors

Universal Wireless Personal Communications, Ramjee Prasad

WCDMA: Towards IP Mobility and Mobile Internet, Tero Ojanperä and Ramjee Prasad, editors

Wireless Communications in Developing Countries: Cellular and Satellite Systems, Rachael E. Schwartz

Wireless Intelligent Networking, Gerry Christensen, Paul G. Florack, and Robert Duncan

Wireless LAN Standards and Applications, Asunción Santamaría and Francisco J. López-Hernández, editors

Wireless Technician's Handbook, Second Edition, Andrew Miceli

For further information on these and other Artech House titles, including previously considered out-of-print books now available through our In-Print-Forever® (IPF®) program, contact:

Artech House
685 Canton Street
Norwood, MA 02062
Phone: 781-769-9750
Fax: 781-769-6334
e-mail: artech@artechhouse.com

Artech House
46 Gillingham Street
London SW1V 1AH UK
Phone: +44 (0)20 7596-8750
Fax: +44 (0)20 7630-0166
e-mail: artech-uk@artechhouse.com

Find us on the World Wide Web at:
www.artechhouse.com